The Chocolate Tree

Smithsonian Nature Books

Books in this series describe the biology and natural history of individual plant or animal species, and offer insight into how science is conducted. Written by established experts in clear, jargon-free prose, the books are intended for naturalists and biologists alike.

THE BALD EAGLE
Haunts and Habits of a Wilderness Monarch
Jon M. Gerrard and Gary R. Bortolotti

HARRIER, HAWK OF THE MARSHES
The Hawk That Is Ruled by a Mouse
Frances Hamerstrom

THE NORTH AMERICAN PORCUPINE
Uldis Roze

RED FOX
The Catlike Canine
J. David Henry

WHITE IBIS
Wetland Wanderer
Keith L. Bildstein

The Chocolate Tree

A Natural History of Cacao

Allen M. Young

Smithsonian Institution Press

Washington and London

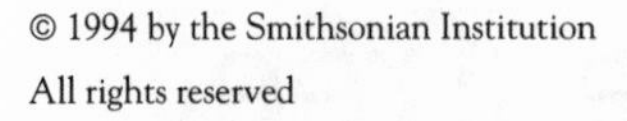

Copy Editor: Tom Ireland
Production Editor: Duke Johns
Designer: Alan Carter

Library of Congress Cataloging-in-Publication Data
Young, Allen M.
The chocolate tree : a natural history of cacao / Allen M. Young.
p. cm.
Includes bibliographical references and index.
ISBN 1-56098-357-4 (alk. paper)
1. Cacao—Central America. 2. Cacao. 3. Cocoa—Central America.
4. Cocoa. I. Title.
SB268.C35Y68 1994
633.7'4'0972—dc20 93-44196

British Library Cataloguing-in-Publication Data is available

Manufactured in the United States of America
01 00 99 98 97 96 95 94 5 4 3 2 1

♾ The paper used in this publication meets the minimum requirements of the American National Standard for Permanence of Paper for Printed Library Materials Z39.48-1984.

This book is dedicated to the many tropical peoples past and present, especially in Mexico, Central America, and South America, who cultivate the cacao tree. It is also dedicated to my friend and colleague, J. Robert Hunter, who helped me develop a research program with cacao in Costa Rica spanning many fruitful years.

Contents

Preface ix

Acknowledgments xiii

1.
Out of the Rain Forest: The Journey of Chocolate Begins 1

2.
The Cultivation of Cacao Past and Present 14

3.
Cacao and Agriculture in Costa Rica 48

4.
Excursions into the Natural History of Cacao and Cacao Plantations 80

5.
Nature in the Cacao: Mysteries of Pollination 107

6.
Back to the Rain Forest: A Bridge between Agriculture and Conservation 155

Appendix. Names of Plants and Animals Mentioned in the Text 175

Bibliography 179

Index 192

Preface

Somewhere in Central America, a thousand or more years ago, an Aztec Indian picks the odd, football-shaped fruit jutting from the trunk and branches of a smooth-barked tree of the rain forest. Perhaps the fruit, encased in a hard, fibrous pod, is a bit past its prime—the normally refreshing white pulp slightly fermented, and the almondlike seeds, or beans, dried out. Perhaps the Indian spits the seeds, or tosses the entire fruit, into a cook fire. As the beans roast, he is riveted by the signature aroma we now associate with hot fudge, simmering cocoa, freshly baked brownies, or a newly opened box of Switzerland's finest. Thus might have been born humankind's millennia-old love affair with chocolate.

The circumstances surrounding the discovery of the tantalizing flavor to be had from the cocoa bean remain a mystery, as do many historical and biological aspects of cacao, the tree that produces the beans. This book delves into the mysteries as it examines the natural history of cacao (Chapters 1 and 4) and its transformation into a cultivated crop of ancient and modern peoples (Chapters 2 and 3). Ever present are its ecological connections to the rain forest, where cacao evolved as an understory tree. Indeed, of all the products to emerge from the diversity of the tropical rain forest, none approaches the universal appeal and popularity of chocolate.

Chocolate is a gustatory bond between past and present peoples. As well, it binds the people of the American tropics, where it originated and was first cultivated, with people of other tropical regions where cacao was subsequently introduced into cultivation, and with the consumer nations of North America, Europe, and the rest of the world. Cacao is among a handful of New World tropical plants that, owing to the Spanish Conquest in the late fifteenth century, became a bridge between two very distinct spheres of humankind: Western culture and society on the one hand, and the ancient and indigenous peoples of Mesoamerica on the other.

Ancient Mesoamericans were both producers and consumers of chocolate. Today, impoverished people in the tropics grow cacao, but chocolate, the processed product, is in large measure synonymous with Western affluence.

Sometimes when a species is domesticated for the production of a commodity such as chocolate, there is a misconception that this dislodged fragment of nature is "ours." But cacao has not been entirely tamed. Within a cacao orchard, created to yield large quantities of cocoa beans, there remain vignettes of tropical rain forest nature, especially where the orchard adjoins the forest. I came to appreciate these glimpses of rain forest nature in the cacao plantations of Costa Rica's Sarapiquí District, in the northeast corner of the Atlantic coast, where I have conducted fieldwork for the past twenty-five years.

Consequently, much of this book is set in Costa Rica, geographically close to the ancient seat of cacao cultivation in southern Mexico and northern Central America. After the long-ago discovery of chocolate, cacao became one of the most important cash crops of the lowland Mesoamerican tropics, shaping the economic, social, and political foundations of ancient peoples. Today Africa is the world's major producer, but the ancient heritage is still evident in Central America, where cacao cultivation continues to be a modest economic determinant.

The interplay of cacao orchards and rain forest on the hills and plains of Costa Rica is vivid and striking. At my two research sites along the windswept slopes of the Cordillera Central in Sarapiquí, the land is a sprawling patchwork of high-canopy, dark green rain forest and low-lying, pale green, sometimes crimson boughs of cacao.

In 1977, I was winding down seven years of fieldwork on cicadas in Sarapiquí and exploring other avenues of potential study. I had seen the cacao groves in the area on many occasions but paid little attention to them. They seemed biologically monotonous compared to the teeming rain forest towering over them. Soon thereafter I came to realize that cacao, harnessed into human cultivation a thousand or more years ago, had a most intriguing biological story to tell—a story that links the tree's natural history with people.

The cacao tree is a testimonial to the complex biological fabric of the tropical rain forest, a weave of ecological specialization and interdependency among plants, animals, and other life forms. The specialized existence and life style of the cacao tree, apparent even in the plantation setting, includes its curious floral design of rigidly separated male and female parts; its peculiar habit of cauliflory, in which the flowers arise from the trunk and primary branches of the tree; and its large, sturdy fruits, called pods, that contain the bitter-tasting beans coated with a delicious mucilaginous pulp. In the rain forest these seeds are dispersed only by monkeys, bats, rats, and squirrels, which gnaw through the pods. That cacao pods do not drop off the tree, even when fully ripe, indicates this plant's extreme ecological dependency upon vertebrate animals to reproduce—a rare if not unique strategy among plants.

On cacao plantations, workers must use machetes to cut the pods from the tree. More often than not, the agronomic yield of beans is small. I was puzzled by the low yields, given cacao's long and widespread history of cultivation and the large number of flowers the tree produces. Were the low yields a result of low levels of pollination?

Upon reviewing the scientific literature on cacao, I was surprised to discover there were several important questions about its pollination—and even its chief pollinator—yet to be answered. For instance, were yields low because pollinators are scarce on plantations? I eventually came to devote a considerable amount of time to studying the pollination of cacao's mystifying flowers. I also began to think about how to enhance natural pollination on Central American cacao plantations in an ecologically and economically sustainable manner. I liked the idea of studying the cacao tree not only as a native species of the American tropical rain forest, but also as an important component in the livelihood of small-scale farmers in the humid lowland tropics.

The conclusions drawn from my research on cacao pollination (Chapter 5) provide the central theme of the book's final chapter (Chapter 6), namely, that successful natural pollination of cacao is linked to the ecology of the tropical rain forest. The ties between cacao and the rain forest bode well for the future of both economic development and biological conservation in the lowland tropics. Based upon my experiences in Costa Rica, these do not have to be opposing forces.

Rather, they can be cooperative programs aimed at helping small-scale farmers rework areas already in cultivation while protecting surrounding rain forest as biological reservoirs needed to stabilize and ensure successful agriculture.

The story of cacao encapsulates the most challenging issue facing the lowland humid or wet tropics today: how to achieve sustainable economic development while protecting what little natural rain forest is still left. I hope that my focused experience with cacao pollination will help illuminate how we might achieve this needed balance between agriculture and biological conservation.

Acknowledgments

My first opportunity to explore a cacao grove was at Finca La Selva, bordering the Río Puerto Viejo in Sarapiquí, during October 1968, when I was associated with the Organization for Tropical Studies in Costa Rica. Subsequently, J. Robert Hunter introduced me to the mysteries of *Theobroma cacao* natural history and agriculture on his farm, La Tirimbina, in Sarapiquí, not far from La Selva.

I was generously aided in my studies of cacao pollination by many people over several years. In addition to Bob Hunter, the cacao specialists at the Centro Agronómico Tropical de Investigaciones y Enseñanza in Turrialba, especially Gustavo A. Enríquez and José A. Galindo, assisted with my logistical needs for conducting research both at that institute and at Finca Experimental La Lola. Especially helpful to my endeavors to study cacao in Costa Rica's Atlantic zone were the friendships I developed with many people who tended the cacao groves.

The American Cocoa Research Institute (ACRI) of the Chocolate Manufacturers Association of America has generously supported my cacao pollination studies continuously from 1978 to the present. I am very grateful to the staff of

ACRI for their interest and support, especially Richard T. O'Connell, president; Russell E. Larson and L. H. "Hank" Purdy, former scientific advisors; and Rhona S. Applebaum, current vice president for science. I owe special thanks also to many others associated with ACRI for their encouragements, especially B. K. Matlick, Gordon R. Patterson, and Glenn A. Trout of Hershey Foods Corporation; Edmond Opler, Sr., of World's Finest Chocolate Company; Kenneth Manning and Ed Minson of Ambrosia Chocolate Company; and Robert Fulton as well as Bob Hunter.

My thanks are extended also to M. Kenneth Starr and Barry H. Rosen, respectively, former and present directors at the Milwaukee Public Museum, for their helpful perspective on the role of research in the quality development of the Milwaukee Public Museum. Several colleagues at the Systematic Entomology Laboratory, U.S. Department of Agriculture, National Museum of Natural History, deserve my sincere thanks for identifying midges collected in my cacao pollination studies over many years, especially Raymond J. Gagné, Willis W. Wirth, and F. Christian Thompson. Barbara and Eric H. Erickson, Jr., David Severson, and others at the University of Wisconsin, Madison, generously added a dimension of high quality to my research on cacao pollination through their collaboration on both field and laboratory studies of floral biology and attractants. I thank William Fowler of Vanderbilt University for very helpful assistance in understanding the role of cacao in pre-Columbian Central American societies and culture.

I am very grateful to the dedicated field assistance given to my cacao pollination studies in Costa Rica by many helpful individuals, including Melanie Strand, Jesús Sánchez, Jorge Mejías, Miguel Cerdas, Ricardo Palacias, Alfredo Paredes, Charles Hunter, Barbara and Eric Erickson, Bob Hunter, Dave Severson, Raymond P. Guries, Mary Joan Treis, Heidi Landecker, Gustavo Enríquez, José Galindo, Victor Villalobos, Arnold Erickson, and Mary Dykstra.

I thank Eric H. Erickson, Jr., William R. Fowler, J. Robert Hunter, and two anonymous reviewers for reviewing parts or all of the manuscript, although they are not responsible for the final contents.

My acquisitions editor at the Smithsonian Institution Press, Peter F. Cannell, encouraged me to write this book. His enthusiasm made my task much easier, and very delightful. Chris Coradini, with her usual professionalism and dedication, did an excellent job of typing various drafts of the manuscript. I thank Christine Mlot and Tom Ireland, with support from the Smithsonian Institution Press, for extremely helpful developmental editing and copy editing, respectively, of the entire manuscript. The author also appreciates the very generous support for the book's color illustrations received from Messrs. Edmond Opler, Sr. and Jr., World's Finest Chocolate Company.

To all of these people, and others, my heartfelt thanks for making this book possible. Finally, thanks to all of the world's chocolate lovers, who added a strong blush of relevance to my research and, thus, to the genesis of this book.

Out of the Rain Forest

The Journey of Chocolate Begins

With big pods jutting from its trunk and lower branches, the cacao tree would certainly have attracted the curiosity of early peoples living in the rain forest. The overall appearance of the cacao tree—its multistemmed, blackish trunks reaching toward the top of the rain forest understory; its pale, button-sized flowers popping out from the trunks; and those oddly engineered pods containing the yet-to-be prized seeds—must have surely have set these people wondering. What is it? Can I eat its fruit? Is it good for anything else?

The inquisitive nature of human beings long ago brought the cacao tree out of the tropical rain forest and into human history. As one of the oldest known cultivars from the American tropical rain forests, the cacao tree, *Theobroma cacao,* is an ancient bridge between natural and human history. As well, it reminds us that only a minuscule portion of the tropical rain forest's botanical wealth, the immense store of biologically active substances mediating the interrelationships between plants and animals, has been tapped by human beings for practical uses.

When human penetration into the lowland rain forests of the Amazon Basin began between 10,000 and 2000 B.P., these ancient peoples must have been at-

Theobroma cacao. This drawing originally appeared in *Rerum medicarum Novae Hispaniae thesaurus* (1651), by F. Hernández. From "Amazonian Cultigens and Their Northward and Westward Migrations in Pre-Columbian Times," by R. E. Schultes (1984).

tracted to the ripe yellow or red cacao pods. Undoubtedly, they watched various wild animals open the pods and feed on the tasty mucilaginous pulp encasing the seeds. Because humans are inquisitive by nature, these ancient peoples undoubtedly opened the pods, tasted the pulp, and discarded the bitter fresh seeds. Later, about two thousand years ago, viable pods were transported from South America into Central America. There, placed under agriculture, these introduced plants gave rise to a distinctive Central American form of cacao, called Central American *criollo*.

The Origin of Cacao

We know virtually nothing about the origin of *Theobroma cacao*. Our knowledge of the cacao tree stems largely from its status as a cultivated crop, rather than as a wild rain forest tree. Yet the evolutionary history of the cacao tree in its wild state is much greater than the history of its earliest cultivation. Genetically, *Theobroma cacao* is a creature of the tropical rain forest in the context of its reproductive behavior and other features of its life cycle and natural history.

In its natural, wild state, cacao is a member of the lower story of the rain forest. Its distribution, as with any tropical rain forest tree species, is a function of habitat and seed dispersal. The spatial separation of populations within a species often determines the extent of morphological variation in seeds and fruit. Cacao grows best under conditions of minimal fluctuations in atmospheric moisture, and thus in well-shaded habitats protected from strong winds (Hardy 1935). The spatial distribution of wild cacao trees is variable, ranging from single, scattered individuals to large clumps. Expansion of a clump of wild cacao is apparently accomplished by animals dropping the seeds after extricating them from the pods. On wild trees in the Amazon, pods range in color from pale to dark green, sometimes splashed with patches of reddish tinge, with considerable variation in the shape and surface texture of pods from different areas within this vast river system (Pound 1932a).

From these observations, it appears as if wild cacao populations, within or near the center of diversity of *Theobroma*, contain a considerable amount of genetic diversity. Morphological variation, such as cacao pod variations, often reflects genetic differentiation. Because wild cacao usually occurs as scattered small clumps, these relatively isolated subpopulations differentially adapt to highly localized differences in habitat, producing hugely variable populations. This, together with underlying physiological variations, also determined by genetics, shapes the broad range of morphological variation noted in wild populations of cacao in Amazonia.

It is well known that lands drained by the Orinoco and Amazon river systems have been the source, down through ancient history and into recent times, of a great many foods, medicines, and other materials useful to humankind, including cacao. Although allegedly wild cacao occurs in southern Central America (Mora Urpi 1958), it is the upper Orinoco River region and the Amazon region that are considered the evolutionary birthplace of this species in its truly wild state (Stone 1984). Its wild state in Central America has generally been viewed as an extension of the species' geographical range from South America (Stone 1984).

While the genus *Theobroma* dates back many millions of years, *T. cacao* could be a relatively recent species, dating back only ten to fifteen thousand years. According to this view, *T. cacao* appeared on the scene with the arrival of the human species in South America. Ancient peoples in Amazonia, for example, might have selectively crossed two existing species, *T. pentagona* and *T. leiocarpa*, to produce *T. cacao*. According to Cuatrecasas (1964), both of these species occur naturally in South America, as well as southern Central America. Cacao in ancient South America, under this view, could have been selected not for the seeds (also called beans), but for the delectable pulp surrounding them. Because all evidence suggests that selection in cacao for seeds arose in Mesoamerica, and since there is evidence that the seed pulp was used as a beverage by South American native peoples, it seems plausible that these people manipulated wild *Theobroma* species to use the pulp. This argument would be even more attractive if indeed the pulp contained similar amounts of theobromine, the kick substance in cacao seeds. Selection for the pulp would appear more adaptable to the humid lowland tropical climate, where *Theobroma* species flourished, because the seeds could not be dried out very easily and would mold over quickly. The later cultivation of cacao in Mexico and Mesoamerica, chiefly in dry regions, could have greatly aided the drying of seeds for use as chocolate. Elaborate sea coastal trade routes flourishing around A.D. 1000 and earlier could have provided the means of transporting cacao seeds from northern South America and southern Central America to Aztec and Mayan centers to the north.

Outward from Amazonia

The extent to which *T. cacao* was dispersed outward from its center of origin in Amazonia by natural means (such as seeds released by animals from the tree's pods and the movement of seeds along streams and rivers) or by prehistoric peoples cannot be ascertained at this time. DNA hybridization studies of differentiated cacao populations could provide a time-dimension framework in which

View inside a typical cacao plantation in Central America. Note the cauliflorous fruits on the tree trunks. From the La Lola Experimental Farm, near Siquirres, Limón Province, Costa Rica.

Theobroma cacao: floral buds, open flowers, and recently set fruit, also called a cherelle.

to assess the comparative roles of cacao dispersal by nature and humankind. But for now, distinguishing between natural and human-influenced distributions remains problematic. Because of the extensive and complicated people-mediated manipulation of *T. cacao*, its origin is largely obliterated, making it difficult to know for sure that what is called *T. cacao* today is really a single, highly variable species or a complex of distinct species.

It appears likely that these two main types of cacao, *criollo* and *forastero*, differentiated from wild populations of cacao in the upper Amazon region (Barrau 1979) and that *criollo* subsequently became the widespread form in Mesoamerica that was domesticated by Indians in that region. *Forastero* cacao may have been subsequently domesticated by Indians in the Amazon-Guiana region (Barrau 1979). Richard Schultes provides an intriguing explanation for the spread of cacao across Amazonia into Central America by prehistoric peoples, based largely upon the examination of the distribution patterns of wild cacao in relation to geography (Schultes 1984). He believes that cacao was first spread throughout the Amazon Basin by people and eventually worked its way northward and westward. Cacao became established at the mouth of the Amazon and then was transported slowly along the humid forests of the Atlantic coast of northern South America, across the Guianas and Venezuela (Myers 1930). From there, cacao would have been dispersed by people bringing seeds to the most northwesterly

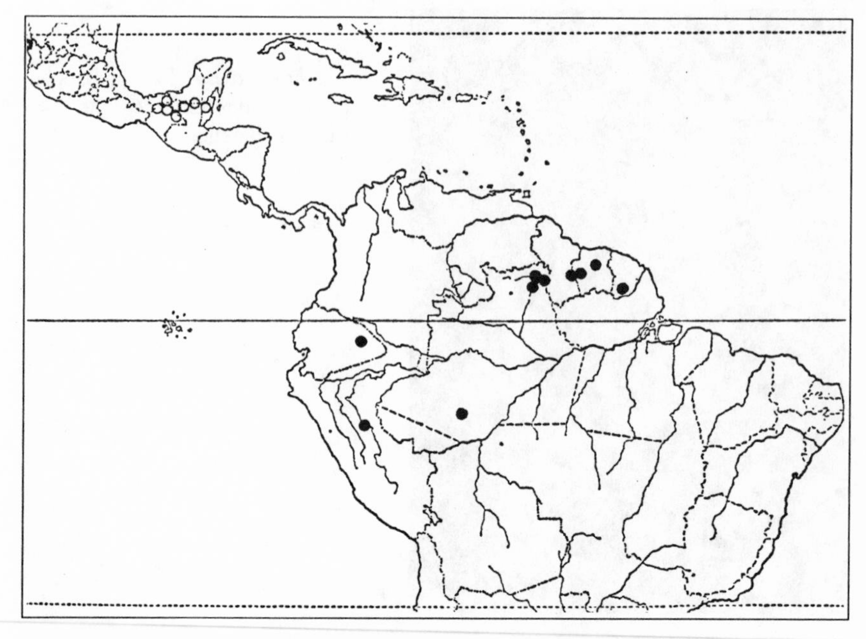

Locations of known, spontaneous, wild populations of *Theobroma cacao* subsp. *cacao* (open circles) and subsp. *sphaerocarpum* (solid circles), which may be the origin of the present cultivated varieties or horticultural races. From Cuatrecasas (1964). Used with the permission of the author.

corner of Colombia, into the extremely wet region of the Gulf of Urabá. Once there, cacao eventually reached Panama, Costa Rica, Nicaragua, Honduras, Guatemala, and southern Mexico, the northern limit of its present distribution. Such a movement pattern would have avoided the barrier of transporting cacao seeds or fruits across the Andes, where cacao seeds most likely would not have survived the cold temperatures (Borroughs and Hunter 1963). But in order to reach the humid region of Colombia's most northwestern corner (Gulf of Urabá), cacao would have had to be carried by people across the arid Caribbean coastal region of this country, because the tree could not survive in this region without irrigation.

Within South America, according to Schultes, cacao might have first been dispersed by way of the Orinoco, where there is a penetration into the Río Negro of Brazil leading to the Río Casiquiare, which functions as a canal linking the Amazon Basin drainage region with the uppermost Orinoco. From here, the required high humidity and rainfall conditions would have facilitated the spread of the cacao tree by people to the coastal region of Venezuela. Although cacao is usually planted below 1,000 meters elevation, it can be grown as high as 4,000 meters in Venezuela, and 3,000 meters in Colombia. It is therefore possible that cacao was carried by Indians from its evolutionary birthplace on the lower, eastern flanks of the Andes (Colombia and Ecuador) through low mountain passes to be planted on the humid, hot western slopes of these mountains. Once established in these humid forest localities of the Pacific coastal region of Ecuador and Colombia, it would have been easy for cacao to then move into Central America and Mexico, without the ecological barrier of an arid region to its dispersal.

The reason ancient peoples in South America disseminated cacao remains a mystery. While ancient peoples may have sucked on the sweet pulp of the seeds, similar pulp is found on many other Amazonian plants. Schultes suggests that some Amazonian tribe may have discovered the use of the caffeine-rich seeds themselves. There is, of course, no written record of this discovery.

It is possible that cacao was carried north to Central America much later, during Preclassic times, the first major agricultural period in Mesoamerican ancient history, from 1500 B.C. to A.D. 200 (Weaver 1981). By the first millennium B.C., extensive long-distance maritime trade routes reached from Ecuador northward as far as Mexico, providing an accessible means of spreading cultural and technical information as well as products like copper and marine shells. Such

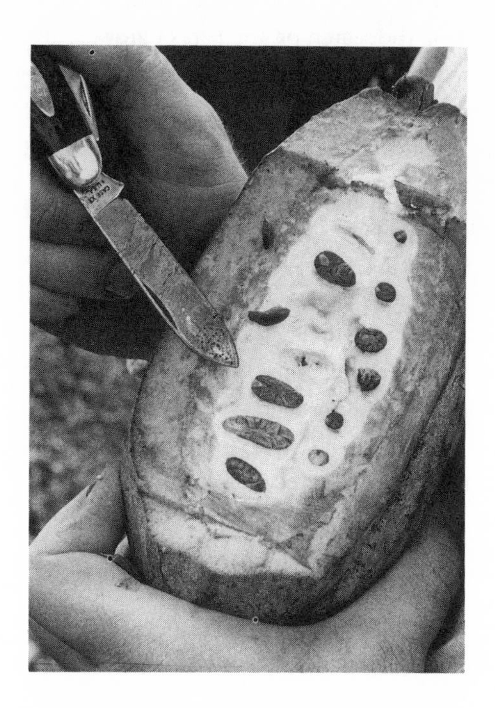

A cacao fruit, cut open lengthwise, revealing the thick outer pod wall, profuse white, mucilaginous pulp, and the dark, developing seeds embedded in the pulp. From the time of pollination, fruits require about six months to develop and ripen, allowing for two harvests a year in many cacao plantations.

a movement of cacao would mean that South American Indian cultures were cultivating cacao on a widespread basis, and much evidence suggests that cultivation did originate in Mesoamerica. This explanation appears less convincing than the earlier, human-mediated spread northwest from the Amazonian drainage region of eastern South America.

In contrast, Cuatrecasas (1964) believed that cacao had a widespread natural distribution in Central America. This is a plausible alternative to the predominantly people-mediated spread of cacao. Allegedly, wild forms of cacao exist in southern Mexico (Chiapas) and Guatemala, which, if true, would mean that cacao had a very widespread distribution in its wild state throughout much of Mesoamerica before its domestication by ancient peoples. Thus, natural events, subsequently modified by human intervention, may also account for the presence of *T. cacao* and other species of *Theobroma* in Central America and Mexico today.

Shaping Cacao's Evolutionary Heritage

Because the cultivation of cacao spread rapidly around the world following European contact with Central America, virtually all cacao crops since have been derived from relatively few wild ancestral types. Thus, from its earliest cultivation to the present, genetic variation in cultivated cacao, as opposed to wild cacao, is presumed to be low, thereby necessitating the search on the part of growers and agronomists for new varieties in nature for selective breeding programs. The degree to which the cacao tree's adaptiveness resides in its genes, or in its somatic variation, or the combination of both shapes the resiliency of this species in its wild state, and as a cultigen.

Collecting expeditions into Amazonia, such as that of the London Cocoa Trade Amazon Project, revealed striking differences among populations associated with different streams and other topographic features within this vast, complex region (Allen 1982; Allen and Lass 1983). How these localized variants came about cannot be known for sure. But the dramatic geological events during the Great Ice Age, or Quaternary, about two million years ago, had a sizable role in shaping the highly diversified floras and faunas of Amazonia. During the first of two epochs comprising the Quaternary—the Pleistocene (which ended about ten thousand years ago) and the Holocene (the present epoch)—cycles of glacial advance and retreat in the northern hemisphere led to corresponding epics of forest expansion and contraction in Amazonia, even though it escaped the direct effects of glaciation, and cycles of arid and humid conditions (Haffer 1982). The repeated episodes of forest fragmentation during the Pleistocene epoch explain in part the biological diversity of the Amazon Basin today, a result of considerable speciation within isolated blocks of forest habitats (Haffer 1969). When forests periodically expanded and formerly isolated areas of forest came into secondary contact, further speciation was set into motion in many groups of organisms.

Under the "refuge theory" (Haffer 1969), isolated breeding populations of plants and animals in constricted forest "refugia," established during the adverse climatic conditions induced by the glaciation cycles, often speciated or at least evolved into different variants or subspecies before a subsequent phase of forest expansion (Haffer 1982). When forest patches expanded, newly differentiated species with developed dispersal abilities spread out over larger areas of habitat before another cycle of forest contraction took place. In this manner, over millennia, much of the rich biological diversity of tropical forests was established. Evidence for the refuge theory comes largely from analyses of the present-day geographical distributions, and habitat associations, of several groups of plants and animals in Amazonian South America (Gentry 1982; Haffer 1982), although it cannot be known whether the refuge theory adequately accounts for the considerable subspeciation of *T. cacao* and other Sterculiaceae within Amazonia.

It is tempting to suggest that during cycles of unfavorable climatic conditions, the expansive rain forest in Amazonia contracted to gallery forest along scattered tributaries. During these periods of isolation, tree species with relatively short-distance pollination systems and long-distance seed dispersal may have experienced considerable subspeciation, becoming fairly isolated variants once the rain forest expanded during subsequent favorable circumstances of climatic change. Perhaps before the Quaternary glaciation two million years ago, *T. cacao* had been uniformly distributed throughout much of Amazonia and was transformed into distinctive variants during periods of forest contraction.

It is also possible that a thousand years or so ago, in the Holocene, Indians moved cacao seeds along Amazonian rivers and streams, since these waterways provided the readiest means of transportation through the rugged terrain and once-extensive forest. Accordingly, the formation of natural variants in cacao would be at least, in part, very recent in the Amazon Basin's complex evolutionary history—on the order of thousands of years, not millions. Such events would have been superimposed in time upon more ancient geological processes, the glaciation cycles of the Great Ice Age (tens of thousands of years ago), which established the high levels of genetic diversity within naturally occurring populations of cacao in Amazonia (assuming an origin in South America). Whether speciation of *Theobroma* and subspeciation in *T. cacao* occurred before, during, or after the Pleistocene and Quaternary refugia has not been adequately studied or analyzed.

Cacao's lengthy journey in space and time, from its likely evolutionary birthplace in South America millions of years ago to its cultivation chiefly in Central America thousands of years ago and the subsequent discovery of chocolate, is consistent with what must also have been true of many other kinds of food or food-flavoring crops in the American tropics. Virtually all discoveries of foods—grains, beans, squash, potatoes, spices, cacao, and so forth—were made many thousands of years ago. Perhaps following the transitions from nomadic to more settled hunting and gathering, and the formation of villages around 5000 B.C. in some parts of South America, people had time to explore garden-style breeding and selection of what were destined to become crops—perhaps more as a leisure-time recreational activity than with a direct intent to discover a new food.

The earliest harvesting of cacao pods from wild populations by Amazonian Indians thousands of years ago was not for chocolate but for the fruit pulp—perhaps mimicking forest animals that would gnaw on the pods to get at it. Indian settlements later in prehistoric Amazonia included gardens of useful plants discovered in the rain forest. In the area of these settlements, extant forests reflect an amalgamation of naturally dispersed and garden-planted tree species. If ancient Indians planted cacao within or near their settlements along streams and rivers in the Amazon Basin to have a ready source of fruit, then allegedly wild trees of this species found in such places today may not be chiefly the result of natural dispersal. From these considerations, one is compelled to wonder just what is identifiable "virgin" rain forest and what was affected by human beings many thousands of years ago. Monkeys, rodents, birds, bats, and human beings undoubtedly interplayed to shape the extant distributional patterns of allegedly wild cacao in Amazonia long ago—and today.

Part of the confusion about recognizing truly wild cacao trees, if they exist as a single species in South American rain forests, comes from the likely movement of cacao fruits and seeds by early peoples along rivers and streams in the Upper Amazonian regions of Ecuador and Colombia, the most likely places to find wild cacao. Early peoples settling along rivers, especially at the confluences of two or more rivers, likely had garden-style plantings of various useful trees, including *T. cacao*, and perhaps other species of *Theobroma* and *Herrania*. Early human disturbance and settlements are most likely to have occurred at river confluences in pristine tropical rain forests. Today, small pockets of what appears to be wild *T. cacao* at river confluences in Upper Amazonia could actually be descendant material from fruits and seeds transported to these places by early peoples thousands of years ago.

In modern-day Amazonia, the Arawete and Asurini Indians occupying the terra firma rain forest region cultivate *Theobroma speciosum*, a distinct species from *T. cacao*, in garden settings, taking advantage of the relatively common occurrence of this species in the tropical rain forest there (Campbell et al. 1986; Balee 1989). Although Indians sometimes make a crude, low-quality chocolate from the seeds of *T. speciosum*, the white, bland pulp surrounding the seeds is more commonly eaten (Cuatrecasas 1964). Similarly, the mature fruits of *Herrania purpurea*, native to Costa Rica and northern South America, are filled with a commonly eaten sweet pulp, although the Bri Bri Indians in Costa Rica also use the seeds to make a bitter-tasting drink (Allen 1956). *Herrania* fruit and seeds might have also been a source of edible oils to early peoples, as suggested by modern-day nutrition analyses of their seeds (MacLean 1952). Collectively and early in the development of Amazonian prehistoric civilizations, the pulp and oils may have been the main refreshment and food commodities provided by the cacao tree.

Not all of the great food discoveries of human beings, most of which had been made many thousands of years ago, were necessarily born out of need. Instead, many of them may have been the result of people having had more time to tinker with and explore fruits in the forest. Without a knowledge of genetics, early peoples in Amazonia, as in other parts of the world, selected and propagated those plant species, such as cacao, that bore seeds and fruits they liked.

Human interest in the cacao tree ultimately centered upon its seeds, but it was the sweet-tasting white pulp—key to the cacao tree's reproductive strategy—that attracted these early peoples. Nutritionally rich, the slippery pulp is the reward for the monkeys, squirrels, rats, and bats that are able to chew open cacao's formidable pods (Janson 1983) and free the otherwise doomed, imprisoned seeds: Unlike the fruit of most trees, even when fully ripe, the mature pods do not fall from the tree. The dozens of pulp-coated seeds themselves are spared from being eaten by animals—the seed cotyledons contain bitter alkaloids including caffeine and theobromines—and eventually give rise to new cacao trees.

It is this same bitterness that in large measure defines that soothing chocolate flavor when the seeds have been first fermented in their own pulp and then gently warmed and roasted—a most mysterious discovery, presumably made early in the course of human civilization in Mesoamerica (e.g., Sauer 1950). Cacao seeds were mashed to make a crude, bitter-tasting paste, mixed with water, chile peppers, vanilla and other spices, and maize to prepare a revered beverage. For this reason, the cacao tree was considered a sacred object in Mesoamerican Indian society. According to Purseglove (1968), cacao came into cultivation in Central America about two thousand years ago from Mexico to southern Costa Rica.

Considered by Purseglove to never have been truly wild in this region, cacao was introduced in prehistoric times from South America. The use of cacao as currency and the belief in its divine origin by early Mesoamerican peoples lend support to the idea that the tree was introduced into the region. It is considered unlikely that a local, wild tree species would have been held in such esteem. This sanctity was recognized and forever sealed in the name given to the tree by Linnaeus: *Theobroma*, "food of the gods."

To the prehistoric Mesoamericans, the cacao tree was the embodiment of the Earth's treasures and spiritually represented a bridge between Earth and the heavens. Perhaps the reddish brown color of this paste reminded these ancient peoples of the earth in which they grew their diversified food crops in gardenlike settings. The ancient peoples of Central America and southern Mexico considered the cacao tree and its special seeds as valuable in their lives as gold and silver, which also came from the earth. Cacao was a gift of nature, offered in rituals as food to

Vertical organization of a typical cacao plantation or grove in Central America today, showing the dense leaf litter beneath well-shaded cacao trees and the overstory layer of legume trees, especially *Erythrina*. From the La Lola Experimental Farm, near Siquirres, Limón Province, Costa Rica.

the gods controlling rainfall, ensuring the success of their crops. Thus, while the tree itself was considered sacred among ancient Mesoamerican peoples, so too, if not more so, were the seeds.

Today, chocolate is the food of romance. People have long thought of it as an aphrodisiac, especially in modern Western societies. Its pleasure is part pharmacological, part psychological, part physical. Chocolate contains more than three hundred identified chemical substances, including theobromine and methylxanthine—two mildly addictive caffeine-like substances—and phenylethylamine, a stimulant chemically similar to the human body's own dopamine and adrenaline. Phenylethylamine acts on the brain's "mood centers" and presumably induces the emotion of falling in love, a matter of only partly understood brain chemistry. Indulgence in chocolate is also viewed as a psychological phenomenon, a craving because of societal pressures to exercise restraint with sweets. Then there is the actual physical pleasure, the feel of chocolate melting in the mouth, that certainly must be part of the seduction. While solid at room temperature, chocolate slowly softens in the mouth, eventually melting just below body temperature.

Only in southern Mexico and Central America did the use of cacao beans for making chocolate take hold before its eventual spread to the rest of the world following European contact with the Aztecs, Mayas, and Pipil-Nicarao peoples in the sixteenth century A.D. Cacao, therefore, has two birthplaces in the New World tropics: its evolutionary origin in South America and its expansive transformation as a cultivated crop in Mexico and Central America. These are two interrelated journeys through different magnitudes of time. The first journey is embedded in the Earth's geological past; the second was driven by humankind in the relatively recent epic of our existence within the past two thousand years.

The Cultivation of Cacao Past and Present

The presence of cacao in Central America is a mystery. Somehow, thousands of years ago, a population of *T. cacao* had spread throughout the central part of Amazonia, westward into Guiana, and northward through Central America to the southern portion of Mexico (Cuatrecasas 1964). Separated by the isthmus of Panama, this cacao population eventually diverged into two distinctive forms, or subspecies: *criollo* cacao in Central America and *forastero* cacao in South America.

Although it has been generally accepted that the Upper Amazon Basin is the evolutionary birthplace of cacao (Pound 1938; Cheesman 1944; Allen 1982), there exists this alternate view that cacao once occurred as a natural population in Central America (Holdridge 1950), highly differentiated from its South American counterpart. It is unclear whether or not these two cacao populations, separated by the Isthmus of Panama, arose from biogeographic (geological) effects according to the refuge theory or are the result of human-mediated dispersal of cacao along trade routes into Central America. The highly differentiated Central American form of cacao *(criollo)* probably arose from spontaneous mutations and subsequent fixing of homozygous recessive phenotypic characteristics in isolated populations (genetic drift) on the isthmus, at the periphery of its natural range (Purseglove 1968).

Human uses of cacao diverged along a similar line. South American Indians sucked on the pulp coating the seeds as a refreshment and used it in a fermented beverage while discarding the seeds altogether (Friede 1953). There is no proof of cultivation or use of cacao seeds in South America in precontact times and, consequently, no evidence of any precontact use of chocolate. Cacao cultivation is absent in the cultural history of the Andean peoples of South America, and the earliest mention of cacao in the region is an account of wild trees in the rain forest of Zaragoza de Antioquia in Colombia in 1636 (Erneholm 1948). Prior to the Conquest, cacao plantations were not found in South America or the Caribbean islands, but only from Mexico to Costa Rica. Within the following three centuries, cacao plantations were established in the Caribbean, South America, and other tropical regions worldwide by the Spaniards, English, French, Dutch, and Portuguese (Morris 1882; Hart 1891, 1911).

Mesoamerican Indians, in contrast, struck upon the process of making chocolate from cacao, which became an important component of Mesoamerican culture and commerce. Cultivation and use of cacao seeds for chocolate and other purposes, therefore, seems to be Mesoamerican in origin. Interestingly, the eastern boundary of Colombia's Choco Province, which runs north-south along the western coast, forms a reasonably discernible point of separation between the two principal ways in which cacao is used by people in Latin America today. Below this imaginary line—that is, throughout much of South America—the fruit pulp is used to make a refreshing, frothy beverage with a citruslike flavor, while the seeds are for the most part discarded.

Above this line, in the Choco and northward throughout Central America and Mexico, cacao beans are used to make chocolate. (This is not to say that indigenous peoples of Central America did not use the pulp of *Theobroma* at all. The use of the seed pulp extends as far north as Nicaragua, where a national beverage, *pinolillo*, is still made today from the pulp of *Theobroma bicolor*.) This dichotomy in the use of the seeds has been in place for millennia. The fact that there are very few names for cacao in the Indian languages of eastern South America, save for *cacau*, suggests that this tree was not of substantial economic importance to the ancient peoples of this region. Other indications come from Humboldt (1884), who noted that wild cacao trees flourished in the Upper Orinoco River region of Venezuela, especially at Puerto Cacao, but that none of the local Indian tribes made a beverage from its seeds. He reported that the Indians sucked the pulp from the cacao pod and tossed the seeds into heaps. Other Indians associated with Spanish missions often collected and then sold these seeds to other people, who made a crude chocolate beverage. Humboldt concluded that chocolate came to South America only with the arrival of the Spaniards, who introduced cacao plantations into Venezuela from their observations of cacao farming in Central America and Mexico. It is interesting to note that cacao plantations had already made a dent in the tropical rain forest of the region to satisfy European commerce by the time of Humboldt's famous expeditions (1799–1804).

The Choco may well represent the entry point of cacao into Central America, most likely prompted by trade routes and the movements of peoples across this region long ago. If the imaginary line is extended further east to the Atlantic, it delineates the most northern portion of Venezuela as another area where cacao beans were used for chocolate rather than the pulp. Interestingly in this context is the fact that the most flavorful types of cacao, the *criollos*, with their starkly white seeds, originated in northern Venezuela and Central America. This observation further suggests the possibility that cacao could have been transported, as ripe pods, by Indians in large canoes along the northernmost coast of Venezuela, across the Caribbean, to places like the southeastern coast of Costa Rica. Therefore, cacao pods could have been taken by seafaring peoples in ancient times from the northern coast of Venezuela, around the Guajira Peninsula of northern Colombia to the west, and into Panama and Costa Rica. It would seem possible that cacao, assuming an origin in South America, could have made its way into Central America by two different Indian trade routes, one along the Pacific, the other across the shoulder between Venezuela and Costa Rica on the Caribbean. While natural dispersal of cacao from South America into Central America cannot be ruled out, the mechanism is very unclear. It is possible that cacao pods could have floated in the sea and reached Central America in this manner.

Long-distance trade routes involving the movement and exchange of cacao beans and possibly seedling trees or cuttings of trees had been well established as early as 1600 B.C., linking Mesoamerica with northern South American peoples (Stone 1984). Whole pods containing fresh seeds, although heavy, could have been readily carried and retained their moisture for as long as two weeks. Because the viability of cacao seeds declines after a few weeks, the spread of cacao cultivation must have been accomplished by vegetative propagation—cacao tree seedlings or cuttings from bigger trees—over long-distance trade routes. The impetus for the spread of cacao cultivation in pre-Columbian Mesoamerica and northern South America might have been fascination with the seeds as a source of addictive, even hallucinogenic substances, of use in popular mystical and ritualistic ceremonies.

Earliest Cultivation

While most scholars agree that cacao was domesticated in Central America and Mexico (Mesoamerica), no one knows for sure exactly when it happened. According to Mesoamerican anthropologist William R. Fowler, Jr. (1989a), cacao was very likely in cultivation in Mesoamerica at the time of Christ, possibly by the Olmec Indians in the Gulf Coast region of Mexico. There is evidence of cacao usage from the first and second centuries A.D. and as far back as 1000 B.C. Possibly as long as two thousand years before the Conquest, wild forms of cacao were already under cultivation by Indians of the Mayoid languages region of Mesoamerica (Bergmann 1969). This early cultivation would have taken place in northern Central America, such as the location of Belize, Guatemala, and Honduras today, and the lowlands of Mexico, such as Yucatán. Much later, around the fifteenth and sixteenth centuries, the Aztecs, a much more recent culture centered in central, highland Mexico, also cultivated cacao. Cacao pods and vinelike images of tree branches bearing pods have figured in aboriginal art, such as Mayan stelae and the more recent Aztec codices (ancient writings). Archaeological evidence from places such as the Nicoya Peninsula in Costa Rica indicate a pulp beverage made from cacao pods was likely used along well-established Mayan trade routes as early as 600–200 B.C. (Stone 1984; Lange 1971).

Jadeite plaque from the cenote of Chichén Itzá, showing a Maya lord grasping a cacao tree while standing on an oversized crab (Terminal Classic period, A.D. 750–900). From Gómez-Pompa et al. (1990). Reproduced with permission of the author and the Society for American Archaeology, as published in *Latin American Antiquity*.

The importance of cacao in the life of ancient peoples is much in evidence from the depiction of the tree and its pods in stylized sculptures associated with Mayan ruins in Central America. Images of warrior-priests and nobles sometimes appeared with cacao trees on stelae or boulders in Mayan territories from the Classic (A.D. 200–900) to Postclassic periods (A.D. 900–1200) (Weaver 1981; Stone 1984). Clusters of cacao pods decorate a stone incense burner dating from the Classic period at Copán (Copán River Valley), in northwestern Honduras, a major agricultural district of the ancient Mayas.

Cacao in Pre-Columbian Mesoamerica

Cacao also appears on tribute lists from many parts of prehistoric Central America—the beans were paid as a tribute to the region's powerful rulers. At the time of the Conquest, in what is now Mexico, cacao beans were offered as tributes to the Aztec emperor from as far away as Soconusco in southeastern Chiapas, the most distant Aztec tributary province (Voorhies 1989). With the expansion of the Aztec Empire from the Valley of Mexico into coastal regions, these Indians demanded tribute of cacao beans from conquered regions. These spoils were brought great distances overland to the Aztec capital of Tenochtitlán in the Valley of Mexico. Soconusco was a prime source of this tribute for the Aztec Empire. According to translations from the *Codex Mendoza* (Clark 1938), cacao was the primary form of tribute from certain regions controlled and populated by the Aztecs in Mexico, and the beans were also used as a form of money. The *Codex Mendoza* was an Aztec book written after the Conquest, and it deals primarily with Aztec daily life, wars, and tribute.

The Aztecs called the tree *cacvaqualhitl*, the harvested fruit or pods *cacvacentli*, the cacao beans *cachoatl*, and the prized drink made from the beans *chocolatl*. In pre-Columbian Mesoamerica, prepared chocolate was not sweetened. In fact, it was often bitter from added chile peppers. (Modern-day Mexican cuisine still employs a mixture of bitter chocolate and chiles in the form of *mole*, a sauce for poultry and fish.) The cacao beans were first dried in the sun, roasted, and then ground on metates (concave, stone grinding platforms) with a little water and other added ingredients, ranging from vanilla (cultivated from an orchid native to Mexico and Central America, *Vanilla planifolia* Jackson) to maize and achiote (the dried red seeds, called annatto or bixin in English, from the bush *Bixa orellana*). The resulting paste was shaped into little cakes and stored until used. The beverage *chocolatl* was made by mixing a piece of a cacao paste cake with water in a gourd vessel; ground achiote gave the drink a reddish color to resemble

Cacao in a cenote, from a painted capstone of the Temple of the Owls (Structure 5C7), an Early Postclassic structure. From Gómez-Pompa et al. (1990). Reproduced with permission of the author and the Society for American Archaeology, as published in *Latin American Antiquity*.

A ripe cacao fruit, split open crosswise to show the thick pod wall and the arrangement of the seeds. The almond-size, reddish-brown seeds are encased in a slippery white pulp.

blood. By tumbling the mixture between gourd vessels, the beverage was prepared to a light, frothy consistency similar to that of whipped honey, making it easy to drink from a vessel such as a finely wrought tortoise shell.

Rev. Thomas Gage, in his eighteenth-century travel writings about Central America (as cited in Dodge 1979) relates the Mayan origin of the term "chocolate" and describes the fruit or pod:

> This name chocolate is an Indian name, and is compounded from *atte,* as some say, or as others, *atle,* which in the Mexican language signifieth 'water,' and from the sound which the water, wherein is put the chocolate, makes, as *choco choco choco,* when it is stirred in a cup by an instrument called a molinet, or *molinillo,* until it bubble and rise unto a froth. But the chief ingredient, without which it cannot be made, is called cacao, a kind of nut or kernel bigger than a great almond which grows upon a tree called the tree of cacao, and ripens in a great husk, wherein sometimes are found more, sometimes less cacaos, sometimes twenty, sometimes thirty, nay forty and above.

This fruit was a valuable crop for the Aztecs. Montezuma, royal monarch of the Aztecs on the central plateau of Mexico, had large storehouses of cacao beans as treasure, not for consumption. Only old, worn beans were used to make *chocolatl*, a ritualistic beverage used at the court of Montezuma at the time of the Conquest (Berdan 1982).

Similarly, cacao was so central to Mayan society that, during the Classic period, the third of the last three rulers at the ancient city of Tikal in Guatemala (the Petén region) was called Lord Cacao. Intensive cacao farming was in place in the swampy jungles of the Petén in Guatemala and in Belize by 250 B.C. (Dahlin 1979), involving large-scale systems of carefully built irrigation canals to drain off excess rainwater during the lengthy rainy season characteristic of this region (Hickling 1961).

Archaeological discoveries attest to the spread of Mayan cacao cultivation from lowlands. In a tomb dating to the Early Classic period, found at Río Azul in the northeast corner of Guatemala in 1984, the lid of a stirrup-handled jar decorated with glyphs (pictographs) bore the statement that the vessel had been used to hold chocolate (Adams 1990). Powdery residues taken from this jar and analyzed by the Hershey Foods Corporation confirmed that it and other vessels found at the site had indeed held a chocolate drink.

A sculptured panel at El Tajin in Vera Cruz, Mexico, a cacao-growing region, depicts a tree with pods on the trunk and branches. Similarly, a tree carved on the Temple of the Wall Panels at Chichén Itzá in the Yucatán may be cacao (Thompson 1956). At the same locality a cacao tree appears on the Teotihuacán mural representing Tlalocán, the paradise of the rain gods. Pottery vessels in the shape of cacao pods have been found in various cacao-growing areas in Mexico, including Cotaxtla in Veracruz, as well as in the cacao region of Guatemala's Pacific coast at Escuintla (see references in Thompson 1956).

As depicted in their art, the ancient peoples of Mesoamerica possessed a sophisticated, insightful perspective on their own existence, one that included a sense of duty to understand the natural world. They understood that they were a part of nature and thus they viewed nature as a friend and partner, rather than enemy. This wisdom extended to the conviction of the Mayas and other prehistoric peoples in the region that it was their responsibility to acknowledge and foster the natural bond linking humankind, nature, and the heavens. Respecting nature was the means to ensure the survival of themselves and future generations. It was largely because of this holistic view of life that these early peoples were successful in developing a diversified agriculture of cacao and other crops within the forests.

The ingenuity and insight of the Mayas in cultivating cacao comes to light when considering the rigorous conditions under which they farmed this crop in the Yucatán of Mexico. Cacao cultivation in Mesoamerica is most extensive in hot and humid regions with total annual rainfall above 2,000 millimeters a year and a short or no clearly defined dry season. Cacao grows well in deep, well-drained soils having lots of organic matter, and usually in association with nitrogen-fixing legume shade trees, especially *Inga*, *Gliricidia*, and *Erythrina*.

Yet a growing body of evidence indicates that ancient cacao flourished in dry areas of Mexico like the Yucatán. Gómez-Pompa et al. (1990) report discovering cacao trees flourishing inside cenotes, or sinkholes, in this region, along with several other tree species of economic importance to the ancient Mayas. Several early chroniclers from Mexico's colonial period referred to the existence of cacao and cenotes in the northern Yucatán as well.

The cenotes, created by groundwater circulating through the karst topography, remain wet and humid despite the dryness of the region. Silt accumulates in the water at the bottom of the cenotes, creating a moist substrate with high air humidity. Since water found in cenotes is not directly affected by rainfall, these places provide a dependable, moist, humid microenvironment suitable for the cultivation of cacao. And since they were scattered, with expansive dry landscape in between, cacao gardens in cenotes were likely to be free of pests and disease organisms, yet highly suitable as a breeding niche for pollinators and other beneficial insects.

Early Varieties

Although three species of *Theobroma* occur in the Mayan region, including *T. angustifolium* and *T. bicolor* in addition to *T. cacao*, the latter has been the most extensively cultivated since ancient times. According to Holdridge (1950), the first type of cacao cultivated in Mesoamerica was called *lagarto* by the Spaniards from what today is Guatemala. A widely held view is that *lagarto*, a *criollo* cacao, was the original cacao cultivated in ancient times in Central America. Mexican *criollo* cacao was originally cultivated in the state of Tabasco and the district of Soconusco in the state of Chiapas down to the border of Guatemala (Bergmann 1969).

Another view (Cuatrecasas 1964) holds that the original wild form of Mesoamerican cacao was Lacandon—a rare form growing as a vinelike tree in the forests in the Lacandon region of Chiapas, Mexico. *T. cacao* subsp. *cacao* form *lacandonense* Cuatrecasas bears small pods with ten rather than five ridges.

The pods of a cacao, *T. pentagonum*, found in the Yucatán cenotes somewhat

Vegetation profile of the cenote Aktun Sitio at Xocen, Yucatán, Mexico. (1) *Ficus yucatanensis*, (2) *Melicoccus bijugatus*, (3) *Sabal yapa*, (4) *Chrysophyllum cainito*, (5) *Brosmium alicastrum*, (6) *Pouteria mammosa*, (7) *Cocos nucifera*, (8) *Mangifera indica*, (9) *Theobroma cacao*, (10) *Citrus sinensis*, (11) *Citrus limonia*, (12) *Musa paradisiaca*, (13) *Xanthosoma yucatanense*, (14) *Bursera simaruba*, (15) unidentified legume, (16) *Annona muricata*. From Gómez-Pompa et al. (1990). Reproduced with permission of the author and the Society for American Archaeology, as published in *Latin American Antiquity*.

resemble *criollo* pods but bear ten rather than five distinct ridges. The same form is found in Chiapas. The Yucatán cacaos are slender trees with primary branches angled upwards, differing from the Lacandon form of Chiapas, which is a climbing tree or shrub ("half-vine"). Other economically useful trees grown in the cenotes by the Mayas included *Citrus*, *Cocos*, *Annona*, and *Musa*, along with nitrogen-fixing tree species.

The discovery of cacao gardens planted by the Mayas in the cenotes of the northern Yucatán peninsula helps in understanding the domestication of this tree in the Mesoamerican region. Trees were selected in the rain forest for transplanting into small garden plots, including those of the cenotes. Seedlings were also cultured in these home gardens. Thus, cacao trees present in cenotes today may be descendants of those cultivated by the ancient Mayas, introduced from rain forest cacao groves in southern Mexico and perhaps even further south. Alternatively, the Yucatán cacao trees in cenotes may represent a distinctive wild form later introduced into southern localities such as Chiapas and the Lacandon rain forest. It is also possible that the tree form characteristic of the Yucatán cenotes and the vinelike form of the Chiapas cacao represent northernmost and southernmost variants of one wild form of *T. cacao* subsp. *cacao* (Gómez-Pompa et al. 1990). Regardless of precisely how the phenotypically wildlike cacao came to the northern Yucatán region of Mexico, it is evident from the cenotes that populations of *T. cacao* subsp. *cacao* were domesticated one or more times in Mesoamerica by ancient peoples before European contact.

In some early accounts from the time of the Spanish Conquest, reference is made to the cacao pod or fruit resembling a "great cucumber but grooved and red" (see Millon 1955) and containing bitter-tasting seeds—a reference to *criollo* cacao *(T. cacao)*. But as many as four kinds of cacao, based on the size of the tree and the beans, were recognized by the Maya Indians. Included in this vernacular classification was a kind of cacao called *tlacachuaquahuitl*, or "humble cacao," coming from small pods and used more for a beverage than for trade (Ciudad-Real 1872; Torquemada 1723). Small prehistoric garden plantings of cacao included more than one variety or kind of cacao, based on the appearance of the pods and beans. Food crops such as *yuca*, a tuber manihot, cassava, or tapioca were sometimes intercropped between rows of young *T. cacao* trees, along with shade trees such as *Theobroma bicolor*, a wild cacao called *pataxte* by the Mayas; and *madre de cacao*, which is the legume tree *Gliricidia sepium*, still used today to shade cacao in places like Belize.

Cacao and the Pipil-Nicarao

All early Mesoamerican portrayals and accounts of cacao indicate a mixture of mystical or religious symbolism along with economic usage. Clues to the economic importance of cacao in the pre-Columbian times of ancient Mesoamerica are provided by the Pipil-Nicarao Indians. These two Indian ethnic groups were derived from a common stock with the Aztecs occupying the Mexican central highlands and the Gulf Coast region.

Before the Spanish Conquest, several waves of migrations carried some of these peoples from the Mexican central highlands into Central America as far south as Nicaragua. Mayan-speaking Indian groups moved into Central America from Mexico, but chiefly into the Petén region of Guatemala, Belize, and Honduras. The greatest movements involved the Pipil and Nicarao ethnic groups. The Pipil settled chiefly in what is today El Salvador, Honduras, and Guatemala, while the Nicarao settled in Nicaragua (Fowler 1989b).

The Pipil-Nicarao peoples migrated into Central America in several phases from A.D. 700 to 1350, an outstanding example of the large-scale migration of

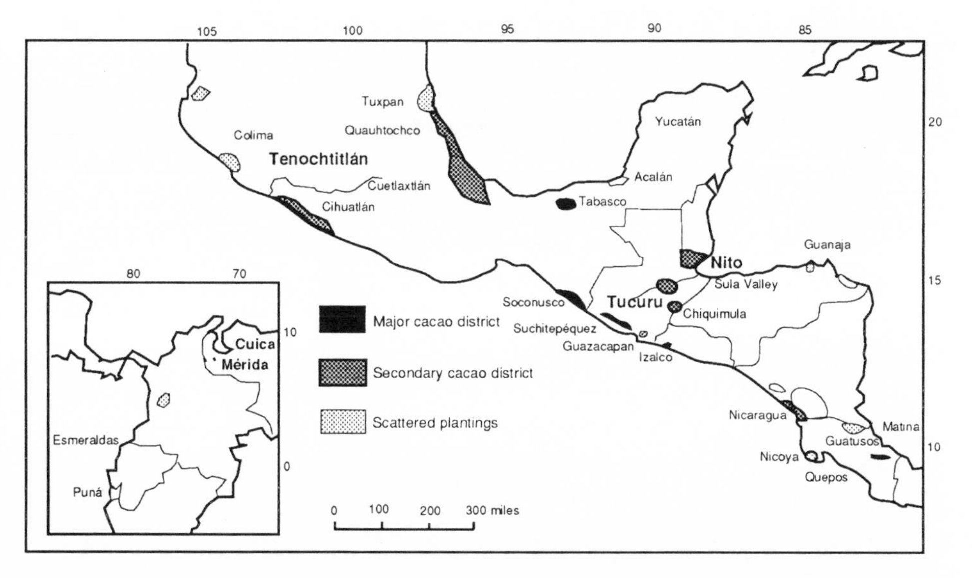

The distribution of cacao cultivation in Pre-Columbian America. From Bergmann (1969). Reproduced with permission from the *Annals of the Association of American Geographers* 59:85–96.

people in Central America's prehistory, and resulting in the eventual movement of the Nicarao into Nicaragua about A.D. 1000 (Borhegyi 1965). The Pipil eventually came to control the good farming lands of Pacific coastal Central America by the time of the Spanish Conquest, with major settlements in the Escuintla region of southeastern Guatemala and most of western and central El Salvador. Pipil settlements in southeastern Guatemala and western-to-central El Salvador encompassed a broad range of elevations, including Pacific piedmont and coastal plains, lowland river basins, upland volcanic basins, and mountain valleys (Fowler 1989c).

The Pipil also controlled the trade routes throughout much of southeastern Mesoamerica before the Conquest, and their broad population distribution reflects the diversity of the commodities they acquired and used to foster a complex economic system. Under the Pipil, different regions of Mesoamerica excelled in specific commodities, encouraging an economic interchange of goods throughout the isthmus. Some of this regional economic specialization focused intensively upon cacao (Fowler 1989c).

The Nicarao settled in northwestern Nicaragua, the Rivas region of southwestern Nicaragua, and very likely part of the Guanacaste region of northwestern Costa Rica. These regions were shared with other Indian groups of Mexican origin, including the Chorotega (Fowler 1985). Costa Rica also became home for

Indian groups originating in South America; its fertile Atlantic or Caribbean region was occupied by Indians belonging to the Chibcha language group, which is centered in Colombia.

The cultivation of cacao was an economically central activity for the Pipil-Nicarao Indians. They developed systems of irrigation to grow the crop in highly seasonal regions of Central America, especially in southern Guatemala and at the major Pipil settlement of Izalco in Pacific coastal El Salvador. These two regions, together with western Nicaragua, despite their extended dry seasons and low rainfall, were among the chief cacao-growing regions of Central America in pre-Conquest times (Fowler 1985).

Because cacao trees cannot tolerate a dry season of four or more months, regardless of the total annual rainfall in a region, the Pipil-Nicarao Indians constructed canals to divert water from permanent streams for irrigating their cacao orchards and circumvent the long dry season associated with Pacific coastal El Salvador. These irrigation canals were therefore vital to the success of their cacao trade.

The Pipil-Nicarao cultivated three species of *Theobroma: T. cacao, T. bicolor,* and *T. angustifolia* (*cacao, pataxte,* and *cushta,* respectively). At least the latter two were endemic to Central America. But only *T. cacao* was of great commercial importance for its use as money and in beverages to the Pipil-Nicarao (Fowler 1985). *Pataxte* was often grown in small cacao groves as a shade cover, together with a tree in the legume or pea family, *Gliricidia sepium,* which added nitrogen to the soil. One gets the impression that the Pipil-Nicarao were indeed expert cacao growers, realizing the importance of adequate soil moisture maintained through shade cover and the effective use of irrigation canals for the same purpose. These people were also successful farmers of many other crops, including cotton, tobacco, coca, maize, beans, tomatoes, avocados, chile peppers, squash and related gourd vegetables, and root crops such as *yuca* and peanuts. The latter were called *cacao de la tierra.*

Cacao was the primary commercial crop of southeastern Mesoamerica from about the Middle Preclassic (1000 B.C.) onward, with a well-established central role in the economy of the region by the Late Preclassic (400 B.C.). It was often traded for products such as salt, bird feathers, animal hides, and obsidian. Cacao is perceived to have mediated the "Mexicanization" of southeastern Mesoamerica as the Pipil-Nicarao peoples moved out of Mexico and through Central America in the Late Classic period (A.D. 700–900 in particular), evidenced in the cacao pods and trees depicted in ceramics and artistic sculptures in Mexico, Guatemala, Honduras, and elsewhere from this period (Fowler 1987). By the time of the Conquest, Izalco, in El Salvador, was the leading center for cacao production and trade.

In addition to being an object of commerce, cacao held a certain symbolism for Mesoamericans. Monuments at the sites of ancient cities in Central America depict cacao pods and trees in various contexts. Some of these have been interpreted in terms of the cacao bean necklaces worn by Pipil Indians during their sacrificial religious rituals, while others bear little or no connection to the matter of sacrifice in these peoples. The cacao pod has also been interpreted as analogous to the human heart—one a vessel for chocolate, the other for blood (Thompson 1956). In some Mayan riddles, chocolate is seen as a symbol for blood (Roys 1943). Great care and ceremony was involved in the planting of the all-important cacao trees. The beans were carefully selected to be planted in accordance with the phases of the moon (García de Palacio 1860, 1881).

What is clearly evident from much of the foregoing discussion is that before the Conquest and the ensuing Colonial period, the Mayas and the Aztec-related Pipil and Nicarao people in Mesoamerica were collectively successful in the development of agricultural systems and elaborate trade routes throughout the region. Virtually all of this would change with the advent of the Conquest and the invasion of Mesoamerica by foreign cultures.

Cacao after the Conquest

With the Conquest came the subjugation of highly successful groups such as the Pipil-Nicarao and the imposition of the *encomienda* system, which altered the history of Central America forever. The *encomienda* was a temporary grant of land from the Spanish Crown to Spanish colonists, the *encomenderos*. The *encomendero* was given a plot of land to exploit, along with the Indians living on it, keeping a fixed percentage of the profits from cacao for himself. Usually after three generations, the land reverted to the Crown. Through this altered trajectory, cacao would still play a central role—not so much within the context of the ways in which native American peoples valued the tree and its seeds, but as a new commodity in the commerce of the world opened up by the Spaniards.

The earliest known Spanish account of cacao cultivation in Mesoamerica comes from the writing of Gonzalo Fernández de Oviedo y Valdes, an official chronicler for the Spanish Crown between 1532 and 1557. In *Historia general y natural de las Indias*, Oviedo describes cacao under cultivation in Nicaragua:

> I first want to describe how they grow and cultivate these trees (cacao) as precious things. They plant in the lands that seem fertile and good and they choose a site with water close by for irrigation during dry periods. They plant them in straight lines and separated 10 to 15 feet in between to allow enough space because they grow and crown out in such a manner that below them all is shaded and the sun cannot reach the earth except for a few parts between some branches. Because some years the sun scalds them in such a way that it fruits in vain and doesn't form correctly and is lost. To remedy this they put other trees in between, the Indians call these other trees *Yaquaquyt* and the Christians call them blackwood. They grow almost twice the size of the cacao and they protect them from the sun, and make shade with their branches and leaves.

This early writer also documents that dried cacao beans were used in sixteenth-century Nicaragua as money. According to Oviedo, a rabbit could be purchased in markets for 10 cacao beans, a horse or mule for 50 beans, and a slave for 100 beans.

Rev. Thomas Gage, in his extensive writings, especially *Travels in Central America* (originally published in 1775 but republished in 1958; and as cited in Dodge 1979), is credited with the first English-language description of the cacao tree. Gage alludes to a wild species of *Theobroma* that he observed was sometimes planted in association with cacao trees by the Indians:

> The tree which doth bear this fruit is so delicate, and the earth where it groweth so extreme hot, that to keep the tree from being consumed by the sun, they first plant other trees which they calls "mothers of the cacao" and when these are grown up to a good height fit to shade the cacao trees, then they plant the *cacauatales* or the trees (orchards) of cacaos, that when they first show themselves above the ground, those trees which are already grown may shelter them, and, as mothers, nourish, defend and shadow them from the sun. The fruit doth not grow naked, but many of them, as I have said, are in one great husk or pod, and therein besides, every grain is closed up in a white juicy skin which the women also love to suck off from the cacao, finding it cool, and in the mouth dissolving into water. There are two sorts of cacao. The one is common, which is of a dark color inclining toward red, being round and picked at the ends; the other is broader and bigger and flatter and not so round, which they call *patlaxi* [*Theobroma bicolor*] and this is white and more drying, and is sold a great deal cheaper than the former.

Europeans had different impressions of the appeal of the *chocolatl* beverage (Humboldt 1884), discovered by the Spaniards around 1530. One described the crude, bitter beverage as "fitter for hogs than men." Hernando Cortés, the explorer, and his page praised the Indians' chocolate as a nutritious, refreshing drink. "He who has drunk one cup," according to the page, "can travel a whole day without any further food, especially in very hot climates" (Humboldt 1884). Despite other descriptions of its black frothy appearance as nothing short of disgusting, the Spaniards nonetheless consumed *chocolatl* to great excess.

By working the cacao orchards under the *encomienda* system, the Indians become a pool of forced labor that provided Spain with a newly found chocolate wealth. Thus, Indians in Central America were forced into the global economic system of the sixteenth century, which eventually destroyed Indian societies and cultures.

The original *conquistadores* and settlers were rewarded by the Crown for their services with *encomiendas*. The best *encomiendas* in Guatemala were given to the family members and friends of the first president of the Audiencia (the governing body of a Spanish governmental district in Conquest times), Alonso Maldonado. Maldonado grantees extorted Indian cacao producers through the use of the tribute system (MacLeod 1973). The more cacao orchards an Indian owned, the more tribute was assessed to him. But tribute to the Spanish lords meant beans for chocolate, and not currency as it did for the Indians (who also used cacao for chocolate). Thus, tribute under the Spanish rule was really a quota system by which cacao growers provided Spain with an assured supply of the prized seeds. The inheritance or purchase of orchards also meant that the new owner assumed the tribute attached to the property. Thus, even though Indian settlements, including the Pipil's Izalco, grew quite large, much of the profit was siphoned off as tribute to the grantees, resulting in a high level of poverty among the Indians. Only the wealthy nobility could afford to consume *chocolatl*, since doing so was actually drinking money.

Even when the Indian population declined greatly from disease epidemics (1576–77), the *encomenderos* continued to levy the same tribute specified in the 1548–51 tax assessment, with the effect that the survivors paid for the dead! Widows were held fully responsible for tribute from cacao orchards owned by their late husbands, and failure to pay the specified tribute resulted in seizure of the property, which was then turned over to someone else to run. *Encomenderos* hired Indian inspectors to monitor cacao harvests in the orchards and, later, to steal the tribute in beans from the drying patios. Such practices continued despite the denunciation in 1603 by the bishop of Guatemala of excessive and unfair tributes. They stopped only with the eventual decline and death of the cacao industry in Mesoamerica under the *encomienda* system.

Where Cacao Grew

At the time of the Conquest, the Spaniards encountered carefully attended cacao "orchards" distributed widely throughout Mexico and Central America, a testimonial to the Mayas and the Pipil-Nicarao peoples, and their agricultural abilities. These orchards extended from the lowlands of Colima and Tuxpan in Mexico southward into Costa Rica, but only a few were the large producers of the beans that were traded with the Mexican Plateau (Bergmann 1969).

The Spaniards discovered that cacao cultivation was especially extensive in the coastal region of the Gulf of Mexico, the southwest Pacific coast of Mexico, and portions of Guatemala, El Salvador, and Costa Rica (see reference to A. L. de Cerrato, *Tasaciones de tributos* [1548–51], in Bergmann 1969). El Salvador, especially Izalco in the Sonsonate Valley in the west, remained one of the largest centers of cacao production under the Pipil in the Late Postclassic period, through A.D. 1524. Cacao cultivation was so intense there during the sixteenth century that, according to William Fowler (1989a), Spanish archival documents of the time relate in considerable detail the water-rights disputes between neighboring orchard owners over access to irrigation canals in Izalco.

Cacao cultivation is absent today in the Izalco region of El Salvador, but Fowler has been studying the irrigation canals dating back to the Colonial period. According to him, many disputes over orchard ownership, governance, and irrigation rights erupted in the region between the Indians and the Spaniards following the Conquest. Pipil governors charged that black slaves of Spanish lords destroyed their irrigation canals and that Spanish leaders often gave away Indian cacao orchards, passed down through many generations, to other Spaniards.

Izalco became one of the richest jewels in the Spanish Crown because of cacao, ensuring that records of the region's history would be well documented and preserved in the Spanish colonial archives in Seville, Spain. These documents showed that cacao production in Izalco remained in the hands of the Pipil, who successfully worked the cacao orchards not only to pay their tribute but to have enough yield to pay for labor and trade in the market place for other goods. Typically, throughout Central America's Colonial period, the cultures of Indian commodity producers and Spanish merchants blended under the trade economy. This resulted in a mixed, or mestizo, population in many places, although the Indians in Izalco resisted such change. Even today Izalco remains an outpost of traditional indigenous culture in El Salvador.

The cacao orchards in Izalco were divided into hundreds of small parcels owned by Indian families, a pattern that can be traced to pre-Conquest times. Some Indian producers with moderate-size cacao orchards (more than two thousand trees) imported wage labor from the highlands, especially from the Alta Verapaz region of Guatemala, until this labor source disappeared in the 1570s. This highlands labor force was wiped out by being overworked and not adapting well to tropical lowland conditions, leading to rampant disease. Ladinos (people of mixed Indian and Spanish blood) and mulattos, in addition to the Pipil Indians, also had small cacao orchards, as did some Spanish *encomenderos*.

The careful studies of William Fowler (1989b, 1989c) provide insight into the central role cacao played in structuring Pipil society in El Salvador during the Colonial period. According to the Seville archives studied by Fowler, a Pipil household in Izalco occasionally owned or controlled as many as twenty-nine cacao orchards, or 33,570 trees. At their prime these trees might have yielded as much as 400,000 cacao beans, or about 850 pounds, annually. The basic Indian unit of measure was the *xiquipil*, or 8,000 beans. Three *xiquipiles*, or 24,000 beans, was the most one person could carry, called a *carga* by the Spaniards. The total annual tribute levied against a large, successful household producing fifty *xiquipiles* would have been twenty *xiquipiles*, still allowing a surplus to be used for labor and trading.

Towns such as Iztapa and Amayuca in Guatemala at the time paid sizable amounts of tribute, such as 1,200 or 1,300 pounds of cacao beans, as indicated on the annual tax registers between 1548 and 1551 (Bergmann 1969). In many small towns, the cacao orchards were tended by Indians, and the harvested beans were destined chiefly for local use.

Whereas most of Mesoamerica during the Conquest adhered to the Indian units of measurement in defining cacao tribute, in Nicaragua Spanish units of measurement applied. There, units such as *fanegas* and *almudes* prevailed—most likely because Nicaragua, unlike much of the rest of Mesoamerica in pre-Conquest times, did not engage in the commercial trading of cacao; rather, it was used locally as money. Indeed, Oviedo wrote that in sixteenth-century Nicaragua "everything is bought cheap with cacao, however expensive or cheap, such as gold, slaves, clothing, things to eat and everything else. . . . There are public women . . . who yield themselves to whomever they like for ten cacao beans . . . which is their money."

The use of cacao beans as currency had certain advantages, even if they lacked a fixed value. Single beans were low enough in value to allow for small transactions (a few beans could be exchanged for corn). They were durable, able to withstand handling, and they could be stored for years without spoilage.

Some large cacao-producing Pipil towns such as Otasingo in Guatemala were assessed a large tribute of 140 *xiquipiles* a year (2,300–2,400 pounds of cacao) according to the tax records of 1548–51, to be paid collectively by their cacao growers. This sum was owed by eight payers for their cacao orchards (Bergmann 1969). Tribute assessment was allocated by "cacao districts," each consisting of several cacao towns, or *pueblos*. For example, the conglomerate cacao district east of the Río Michatoya in Guatemala, consisting of the pueblos Atiquipaque, Taxisco, Guazacapan, and Chiquimulilla, was assessed an annual tribute of 1,946 *xiquipiles* of cacao beans, or 38,920 pounds. Archival accounts of sixteenth-century Guatemala and El Salvador reveal a combined annual tribute of 15,709 *xiquipiles* (314,180 pounds). One account of Guatemala's annual cacao harvest at the time is six million pounds of beans, making the annual tribute about 5 percent of the total yield. This probably represented a maximum level of cacao bean taxation, since the industry declined sharply in Mesoamerica in the seventeenth century (Bergmann 1969).

At the time of the Conquest, modern-day El Salvador and Guatemala were the regions producing the most cacao in Mesoamerica and the world. Aboriginal plantings of cacao trees yielded as many as 100 harvestable pods each year. On a per acre basis, these ancient yields were as low as the lowest average yields per acre in the 1950s in Central America (Millon 1955).

Izalco experienced a cacao boom in the sixteenth century that, according to Fowler (1985), had two major demographic effects in El Salvador. The high demand for cacao coupled with a large decimation of the Pipil population by plagues (1545–48 and 1576–77), resulted in a large influx of laborers from highland Guatemala until labor importation became prohibited in the 1570s by royal laws. Second, a large concentration of Spanish merchants developed in the town of Sonsonate, founded in 1552, near Izalco. Sonsonate became a powerful center for trading in cacao, where none of the merchants were *encomenderos*. From the beginning, it was a town of only merchants and traders, and no soldiers. This was quite different from the situation in other parts of the Audiencia of Guatemala, where *encomenderos* focused their efforts on monocultural agriculture, especially cacao, in the absence of more lucrative exploration for precious metals. Thus the *encomenderos* of the Sonsonate region (Izalco, Caluco, Naolingo, Tacuscalco) were just as heavily dependent on cacao monocropping as any of the others in the region.

The dependency of the *encomienda* system upon cacao and other monocultural crops under the Spanish system represented what MacLeod (1973) called the *produit moteur*—a product that yields very high profits on a very low or small capital investment generated by cheap or free labor. Under this arrangement, which dominated much of the first half of Central America's colonial period, the *encomienda* system resulted in a select few becoming rich quickly.

Another result was a wide difference in the agricultural and commercial systems of pre-Conquest and colonial Central America. The Indians had developed a highly successful system of diversified agriculture, which largely sustained an elaborate trade system. Much of this system was erased upon European contact, replaced with monocultural agriculture and a world-based commerce that dominated the region in the colonial period and through modern times. Under the world economy of the sixteenth century, part of the Western Hemisphere became incorporated into a system in which raw commodities such as cacao beans were channeled through European core markets, forcing the agricultural system of the New World tropics to respond to the demands of the global economy rather than merely serving its own people and commerce (Fowler 1987).

Spanish trade brought into Central America new commodities from Europe, including glass-bead necklaces and bracelets, brass needles, pins, silver jewelry, olive jars, locks, and nails. These items entered the region as cacao beans and other agricultural commodities went out. In places where cacao was the chief means of tribute, Indians were able to obtain a greater range of European goods than where natives did not use cacao as tribute. In areas where cooperative trade networks existed between the Spaniards and the Indians, a greater level of European culture and religion (Catholicism) was incorporated into Indian culture.

In the mid-to-late sixteenth century, cacao production in Central America declined, the logical consequence of a severely diminished labor pool owing to plagues. Further, the great pressures put upon the Indians by the Spaniards to increase production took its toll on the land, which was exhausted by the beginning of the seventeenth century.

Back to South America

Following the development and expansion of cacao cultivation in Mesoamerica, and then as cacao production began to wane, the Spaniards introduced cacao cultivation into South America. The planting of cacao trees in the form of large plantations in South America thus did not occur until historical times, that is, well after the domestication of cacao in Central America and southern Mexico. In this context, cacao, originally from South America, would not become a cash crop there until after other peoples in Central America discovered the use of the tree's seeds for making chocolate. Only then would cacao become an important agricultural commodity in Brazil, Ecuador, Colombia, and other countries in South America.

The cultivated Central American cacao, *criollo*, was likely taken to Venezuela and the rest of South America by Spanish missionaries and settlers to develop plantations. According to Pittier (1935), in the seventeenth century Capuchin friars (from Spain) brought *criollo* cacao seeds from the Nicoya Peninsula of Costa Rica or from Cuba into the state of Aragua in Venezuela, and from there, cacao spread rapidly into other regions of Venezuela. In the process, *criollo* cacao undoubtedly hybridized with local forms of cacao, giving rise to a complex of types, called *trinitario*, still cultivated today in northeastern South America (León 1984).

Jesuit missionaries in the seventeenth century recorded seeing large groves of wild cacao trees in the Andean foothills of Venezuela. Venezuelan cacao populations might have been a source of material eventually cultivated in the Caribbean, although some of this cacao might have been derived also from Central America. Cacao from Venezuela might have made its way into the Lesser Antilles and the Guianas by European immigrants, possibly reaching as far north as Jamaica, where it might have hybridized with cacao brought from the Yucatán through Cuba and Santo Domingo (i.e., the Dominican Republic) (Pittier 1935).

Cacao also made its way to the islands of the Caribbean with the rise of European colonialism in the New World. This was largely a result of British expansionism and the establishment of the plantation arrangement of cacao trees—neat rows of planted trees encompassing large areas. Earlier, the Spaniards relied chiefly upon the Indian methods of growing cacao in orchards or groves of varying sizes. The cultivation of cacao spread from Central America into colonial Martinique, controlled by the French; Trinidad, under the British; Venezuela, under the Spaniards; and Brazil, under the Portuguese.

From *Chocolatl* to Chocolate

Even though Columbus was the first European to encounter the cacao beverage in Central America, it was not until the sixteenth century, when Hernando Cortés visited the royal Aztec court of Montezuma in Mexico, that the importance of *chocolatl* in the New World was appreciated. In 1519, Cortés witnessed Emperor Montezuma being offered fifty golden bowls filled with *chocolatl*.

Eleven years later, as the captain general of Mexico, Cortés appeared at the royal court in Spain and introduced chocolate for the first time. Because of its bitterness, this crude chocolate was not well received at first. Only when the Spaniards made chocolate more tasteful by mixing a crude cacao paste with sugar and seasoning it with cinnamon—a practice still followed in Mexico today—did it become acceptable and popular. Spanish royalty in the early seventeenth century adulterated cacao with many more ingredients than the Aztecs, who used chiefly vanilla and chile peppers. According to Cadbury (1896), the Spaniards doctored their cacao with chile peppers, anise, achiote, sugar, vanilla, cinnamon, almonds and hazelnuts, pod of campeche, ambergris, powdered white roses, and orange water. In this form, chocolate became a novelty in Europe, the first non-alcoholic stimulant drink on the continent. Coffee did not reach Europe until 1615, and tea arrived there much later. The allure of chocolate to the Spaniards, who kept the prized beverage a secret for almost a century, was its strengthening powers—one could travel all day after drinking just one cup in the morning. Europe's chocolate boom began when Spain lost its hold on the commodity as a result of a royal marriage between Marie Thérèse of Spain and Louis XIV of France in 1660.

A boost was given to chocolate's growing popularity in Europe by the expansion of sugar plantations in Brazil and the Caribbean between 1640 and 1680, slashing the price of sugar and making it more available to be combined with both chocolate and coffee, two very popular beverages in the seventeenth century. Chocolate soon became very popular and in high demand, first in Spain, and later in Italy, Flanders, France, and England. Although merchants may have introduced chocolate to Italy in 1606, Catholic friars engaged in drinking hot chocolate even before this time, no doubt spreading chocolate throughout Europe when they journeyed between various monasteries (West 1992). By the middle of the seventeenth century, sweetened hot chocolate was very popular, especially among the wealthy, throughout much of Europe.

During the seventeenth and eighteenth centuries, the consumption of chocolate, believed to cure many ills and inflame passions, in addition to tasting good, was limited. Only the wealthy could pay the high import taxes placed on the beans or cacao paste in countries such as Spain, France, and England. England's upper strata of society enjoyed a chocolate beverage mixed with milk. Samuel Pepys (as cited in Minifie 1980) penned in his diary in 1664 that he was off "to a coffee house to drink jocolatte, very good." In *A Tale of Two Cities*, Charles Dickens portrayed the habit of chocolate drinking in seventeenth-century Europe as a luxury of the idle upper class, and not of the masses. Early eighteenth-century England saw the proliferation of chocolate houses in London, in large measure outstripping the popularity of the previous century's coffee houses for all classes of society. In the eighteenth century, the British mixed brick dust into their chocolate to thicken it. In 1727 an Englishman, Nicholas Sanders, first mixed chocolate with milk to make hot chocolate, which was promoted by British physicians such as Sir Hans Sloane as healthy for both children and adults. The first chocolate manufacturing company in England was in operation by 1730. Begun by Walter Churchman of Bristol, the company was taken over in 1761 by a Quaker, Dr. Joseph Fry. Subsequently, two other Quaker families, the Cadburys and the Rowntrees, established successful chocolate companies in England.

During the nineteenth century, import duties on cacao beans and other commodities from the New World dropped, and consumption of chocolate increased greatly. Soon thereafter, the industrial revolution, which mechanized the production of chocolate, brought consumer prices down, making it a drink for the masses. Cacao bean imports into Britain alone in the 1820s reached a new high of five hundred tons a year. The British navy at the time consumed roughly half of the chocolate produced in England. Sailors on watch drank "navy cocoa," a chocolate-rum hot drink, until after World War II.

It was not until 1779 that chocolate made its way back to the Americas with the establishment of a chocolate factory in Massachusetts, Baker's Chocolate Company in Lower Mills, Dorchester, the oldest in the United States. The mill was originally established on the Neponset River in 1765 by John Hannon. Dr. James Baker became a financial partner in the company after Hannon was lost at sea in the Caribbean on a trading voyage. The company became Walter Baker and Company, Ltd., in 1779. Today, Baker Chocolate is still a familiar and widely used product. The original mill was probably the only U.S. chocolate factory to grind cacao beans by waterpower. Baker extolled what Dr. Fry had done in England, saying that chocolate was an excellent stimulant to the human body. And earlier, Thomas Jefferson had expressed the hope that "the superiority of chocolate, both for health and nourishment, would soon give it the same preference over tea and coffee in America which it has in Spain."

More and more cacao was needed to soothe the sweet tooth of Europe and the United States, prompting the expansion of cacao agriculture into the Caribbean colonies and South America. The original trade monopoly with the New World, controlled by the Spaniards, was broken when the Dutch took over the island of Curaçao in the eighteenth century, opening new trade markets for cacao beans into Europe. Other countries set up their own plantations, such as England's "cocoa walks" in the colonial West Indies.

During the nineteenth century the tremendous popularity of drinking chocolate prompted key innovations in the fledgling chocolate industry. Chocolate manufacturing was vastly improved with the discovery in 1828 by Conrad van Hooten, a Dutch chemist, that the fat could be pressed out of roasted cacao beans, called nibs, to make cocoa powder. The fat, or cocoa butter, could then be used for other purposes. Defatting the beans and adding other ingredients revolutionized the use of cacao in the candy industry. Van Hooten developed a mechanical press to remove some of the cacao bean fat—a major technological breakthrough in the chocolate industry. Separating much of the fat from the bean left behind a flaky powder, cocoa, with only about 25 percent fat content, making the cocoa easier to mix with other ingredients and easier to digest. Even though cocoa butter is one of the most saturated fats known, it has little or no effect on serum cholesterol levels. Thus, much to the good fortune of chocolate aficionados, cocoa butter is a "good" dietary fat, having a fine taste but poorly absorbed by the body. In 1848 the first "eating chocolate" was created by adding cocoa butter and sugar to a paste of the ground cacao beans. In 1875 the Swiss developed the first solid milk chocolate. Thus, together with candymakers Suchard and Peter in Switzerland, van Hooten developed a means of defatting the beans and adding sugar, nuts, and fruits to make a dark eating chocolate.

The Dutch also developed the process of alkalinization, called dutching, in which alkali is added to neutralize the organic acids in the cocoa bean. Dutching darkens the cocoa, makes it milder in flavor, and increases its solubility. Today about 90 percent of all cocoa is dutched. Fry in 1847 and Cadbury Brothers in England in 1849 developed a technique for reconstituting some of the removed cocoa butter, mixing it with the nibs and sugar. This technique allowed chocolate to be readily molded into bars.

Daniel Peter in Switzerland invented the world's first milk chocolate by mixing milk solids with cocoa and sugar in 1876. The Swiss held a monopoly on the production of the highly desirable milk chocolate until Cadbury developed a dairy milk chocolate drink in 1904. The development and consumption of milk chocolate in various forms became the most outstanding feature of the industry in the twentieth century, making chocolate manufacturing the dominant component of the candy industry on a worldwide basis. The chocolate industry reached new heights with the production of flavorings for ice cream and bakery products, refinement of the "smoothness" or "melt-away" sensation of chocolate candy in the mouth, and the production of chocolate coatings for dessert products.

A Pan-Tropical Crop

With refinements in processing chocolate and its diversified use in modern foods and beverages came heightened demand for farming cacao throughout the humid and wet tropics. In the nineteenth and twentieth centuries, cacao became a pantropical cash crop, with some level of production under way in almost all countries of the humid tropics, including Africa and Southeast Asia. One could say that the pan-tropical approach began with the Spaniards, who successfully transferred *criollo* cacao seedlings from Acapulco, Mexico, to Manila in the Philippines in the seventeenth century, where "Java *criollo*" cacao was developed.

To a considerable degree, early and modern cacao cultivation practices in the Philippines parallel methods developed by the Spaniards in sixteenth-century Mexico (West 1992).

From the Philippines, the cultivation of cacao spread to Sulawesi and Java. The first introduction of cacao into Sri Lanka and India came from the east, as there is evidence of *criollo* cacao coming into India from the island of Ambon in the Moluccas, off Indonesia, in 1798. Thus across the equatorial belt of the Earth, the bulk of cacao cultivation in the sixteenth and seventeenth centuries was *criollo*. In the eighteenth century *forastero* cacao was grown first in Ecuador and Brazil, originating as wild cacao in the Amazon Basin. This *forastero* cacao, called *cacao bravo*, became South America's first export cacao. Extensive planting of cacao in the Amazon Basin began in the seventeenth century, when the crop was a small portion of Brazil's initial exportable production of the prized beans. Exports of cacao beans through Pará from the Amazon was 1,000 tons by the end of the eighteenth century, 2,000 by 1820, and 5,000 by 1880.

An important expansion of cacao cultivation in the eighteenth century came with the introduction of an *amelonado* type from the Amazon Basin in Brazil to the state of Bahia in that country. Cacao seeds were apparently brought by a Frenchman, Frederick Warneau, from Pará in 1746 and given to a planter at Cubículo de Almado on the banks of the Río Pardo (Urquhart and Wood 1954). These seeds were possibly derived from *amelonado* cacao originating in the Guianas, a result of the French attempt to break the Spanish agricultural and trade monopolies in cacao. Before expanding their cacao cultivation in Bahia, the French began producing cacao in their own Caribbean possessions of Martinique and Saint Lucia in about 1660. Many decades later Bahia began to become the major cacao-producing region it is today. A rapid upshoot in cacao production in Bahia occurred near the end of the nineteenth century. Subsequently, as the cacao wars escalated, the Dutch transported cacao seedlings from the Philippines to the Dutch East Indies, establishing an experimental cacao garden in Jakarta, Sumatra, by 1778. Dutch-controlled cacao cultivation soon became widespread and successful in Sumatra and Java.

Criollo cacao was introduced into southeast Asia in 1560, transported from Caracas, Venezuela, eastward to the Celebes (Van Hall 1914; Nosti 1953). Java eventually received planting material from either the Philippines or the Celebes. Cacao plantings, presumably *criollo*, were evidently taken from Amboina in the Molluccas to India in 1798 and cultivated in the Tirunelveli District of Madras (Ratnam 1961).

By the beginning of the nineteenth century, cacao cultivation had spread from Central America to South America and some of the Caribbean islands, in addi-

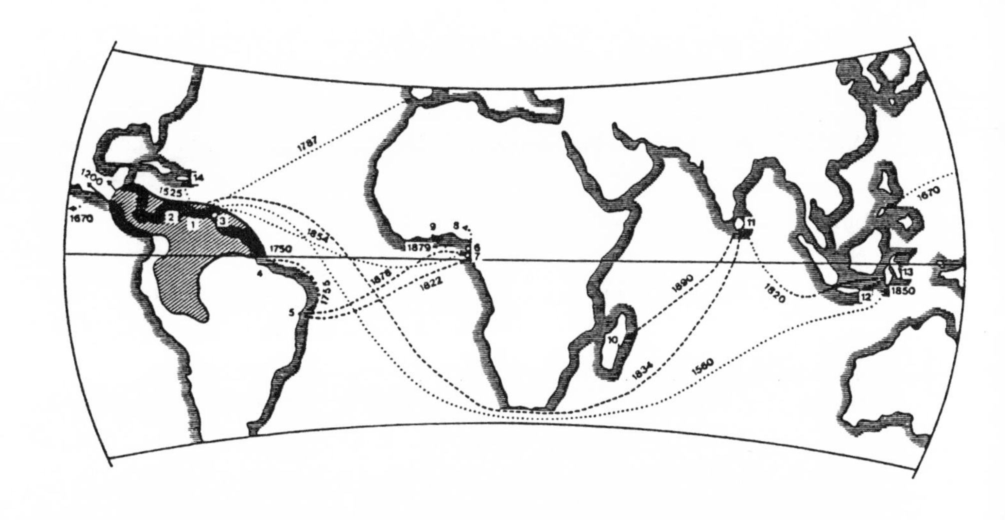

Symbol	Nationality
....	Spanish
------------------	British
————————	Mayas
-.-.-.-.-.-.-.-.-	Portuguese
-..-..-..-..-..-	Dutch

1 Caracas
2 Maracaibo
3 Trinidad
4 State of Pará, Brazil
5 State of Bahia, Brazil
6 Fernando Po (now Bioko)
7 São Tomé
8 Nigeria
9 Côte d'Ivoire
10 Madagascar
11 Ceylon (now Sri Lanka)
12 Java, Indonesia
13 Celebes, Indonesia
14 Dominican Republic

Source:
after Nosti (1953)

Origin and destination of principal movements of cacao, A.D. 1200 to 1879. From Wood (1991). Reproduced with permission of the author and Cadbury Ltd. (Bournville, Birmingham, England).

tion to localized parts of the East Indies. During the nineteenth century there was considerable trade between Brazil and West Africa, resulting in the introduction of cacao cultivation in 1822 to Príncipe, a small island hugging the coast of West Africa. From this island, cacao spread to other islands in the same chain, first to São Tomé in 1830 and to Fernando Po (now called Bioko) in 1854 on the orders of Queen Isabel II of Spain (Nosti 1953). Alternatively, cacao might have been introduced first into Fernando Po by the Spaniards in the seventeenth century from eastern Venezuela (Cook 1982). Cacao cultivation in São Tomé and Fernando Po depended upon workers being brought in from Angola for the former, and from Nigeria for the latter. Cacao was eventually introduced into Nigeria from Fernando Po in 1874 (Ayorinde 1966). Although cacao cultivation never really developed in Nigeria, the introduction of the crop from Fernando Po to the Gold Coast (now Ghana) in 1879 was a much greater success in the expansion of cultivation in Africa. Prior to this, various missionaries had attempted to introduce cacao into West Africa (Wanner 1962), meeting with little success, but paving the way for others.

Amelonado, the type of cacao introduced into West Africa via the Gold Coast, became the dominant type grown throughout the region, with the exception of Cameroon. When Cameroon was a German colony, planting material had been brought from South America and the West Indies, resulting in different *trinitario* cultivars becoming established in Cameroon. For example, the *trinitario* population in West Cameroon had red pods, while those of East Cameroon, having been mixed with *amelonado* from Fernando Po, displayed pods with colors different from those of East Cameroon. Today Cameroon cacao beans possess a higher fat content than the cacao from Ghana and Nigeria, possibly reflecting divergence in the origins of the earlier introductions. A small number of cacao trees of the *criollo* type were introduced in 1920 to Fernando Po, which, when crossed with *amelonado* already established there, gave rise to a highly localized form of *trinitario* cacao (Swarbrick 1956).

The more recent introductions of cacao into West Africa have been Amazon types, initiated when pods were sent from Trinidad to Ghana in 1945 (Toxopeus 1985). These cultivars were developed by the Imperial College of Tropical Agriculture and the Department of Agriculture in Trinidad, based in large part upon material brought to Trinidad from the Amazon Basin to screen for resistance to witches-broom, a fungal disease. Virtually all of the cacao production in West Africa since about 1960 is a direct result of this planting material brought from Trinidad, augmented by ongoing screening of new material at the Cocoa Research Unit at the University of the West Indies.

The introduction of cacao cultivation into Asia and the Pacific rim has taken

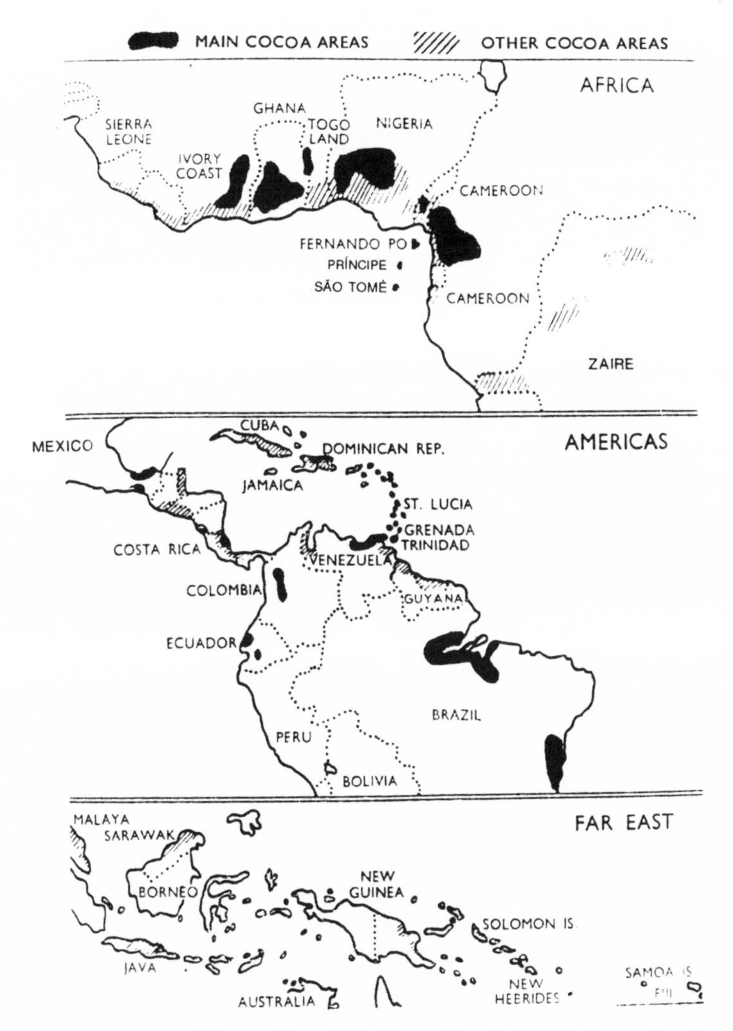

The cacao-growing areas of the world in 1980. From Minifie (1980). Reproduced with permission of Cadbury Ltd. (Bournville, Birmingham, England).

a most circuitous and not well understood route. As mentioned, the earliest introductions into this region were Celebes (Sulawesi, Indonesia) and the Philippines. In the Hawaiian islands, Francisco de Paula Marin planted cacao trees on his estate near Pearl Harbor on Oahu before 1831, when they were observed by the Prussian naturalist Franz Julius Ferdinand Meyen. Visitors from Mexico presumably brought plants of this "Guatemalan cocoa" to Hawaii around this time (Nagata 1985).

An introduction of cacao to Ceylon might have occurred at about the same time. Ceylon in the early nineteenth century became a center for cacao propagation, with garden collections and plantations. There was considerable movement of cacao cultivars, in both directions, between Ceylon and Trinidad during this period. The first introductions from Trinidad took place in 1834 and 1835, and later in 1880 (Wood 1991). By the end of the nineteenth century Ceylon had introduced cacao to Singapore and Fiji (1880), Samoa (1883), Queensland (1886), Bombay and Zanzibar (1887), and to British Central Africa (Tanzania) in 1893 (Wright 1907). Cacao was also sent from Ceylon to Madagascar during this period.

Since cacao seeds lose their viability after less than two weeks, the spread of cacao cultivation throughout the world must have been accomplished by moving young trees or seedlings. The long-distance transport of cacao, often requiring several months at sea, was likely accomplished by carrying the young trees in glass containers known as Wardian cases, miniature greenhouses capable of holding 100 to 150 one-year-old cacao trees (Hinchley Hart 1911). The cases, invented in London for culturing ferns, were extensively used during the nineteenth century (Hinchley Hart 1911; Etter 1971–73). How cacao was moved long distances before this remains a mystery; perhaps other kinds of containers were used for seedlings.

Cacao in the Americas Today

Criollo, *forastero*, and *trinitario* are the three main types or agricultural races of *T. cacao* (Cheesman 1944). These groupings are distinguished from one another by the color and texture of the pods, the color of the beans, and the number of beans per pod (Table 1). *Criollos*, of which there are at least four kinds (Mexican, *lagarto* [also called *pentagona*], Nicaraguan [also called *cacao real*], and Colombian), were probably domesticated by the Mayas and occur chiefly in Central America. *Criollos* were domesticated in Mesoamerica and have red or yellow pointed pods when ripe with thin husks bearing deep furrows and a warty surface texture. The seeds are thick and have white or pale purple cotyledons and a very fine flavor. Central America *criollo*, the oldest kind cultivated in Central America, and the world for that matter, has only weakly astringent seeds requiring very little fermentation to produce chocolate.

Forasteros have thick pod husks and generally flat, dark purple seeds possessing a high astringency. This group is usually divided into Upper Amazonian (as in the wild or semiwild cacao types from this region described in Pound 1938) and Lower Amazonian types, also called *amelonados*. *Amelonados* are the most wide-

Table 1. Main Distinctive Characters of *Criollo*, *Forastero*, and *Trinitario* Cacao

	Criollo	*Forastero*	*Trinitario*
Pod husk			
Texture	soft	hard	mostly hard
Color	red	green	variable
Seeds			
Average number per pod	20–30	30 or more	30 or more
Color of cotyledons	white, ivory, or pale purple	pale to deep purple	variable, white seeds rarely occur

Source: Wood and Lass (1985)

spread cultivated types of cacao in the world today, especially in Brazil and West Africa. *Trinitarios* are intermediate types between *criollo* and *forastero*, which, unlike the latter two groups, have not been found in the wild. These types are derived from *T. cacao* subsp. *cacao*.

At the time of the Conquest the main cacao-producing regions were Izalco in El Salvador and the Sula Valley along the Caribbean Coast of Honduras, although today neither of these areas produces much cacao. Bergmann (1969) wrote that cacao in Conquest Nicaragua was limited to the Pacific coast, chiefly in the districts of León and Granada. Cacao cultivation in Nicaragua has all but disappeared, although some small plantings are found in places like Chinandega. In these places there is a preponderance of *criollos*, more so than elsewhere in Nicaragua. These forms are known as *cundeamor*, with long, bright red pods bearing a constriction in the "neck" and a curved apex or tip; and *angoleta*, with small green pods. In other parts of Nicaragua there are small remnants of *criollos* and pentagonum-type cacao in coffee fields (Soria 1959). In Nicaragua the genetic erosion of the pure *criollo* cacao is about 95 percent today. This means that most forms of pure *criollo* cacao have all but disappeared in extant populations in Central America, having blended with other forms through natural and selective breeding.

In Mexico today only about 20 percent of the extant production is of *criollo* cacao—the rest is of other varieties established in recent times. Small groves of pure *criollo* cacao trees are still found in the state of Tabasco today. Several kinds of *criollo* cacao occur in Mexico today, including *angoleta* and *cundeamor*.

Bergmann (1969) describes the Pacific coast of Guatemala (department of Suchitepequez) near the Mexican border and the foothills of the mountains as important *criollo* cacao-growing districts at the time of the Conquest, as well as

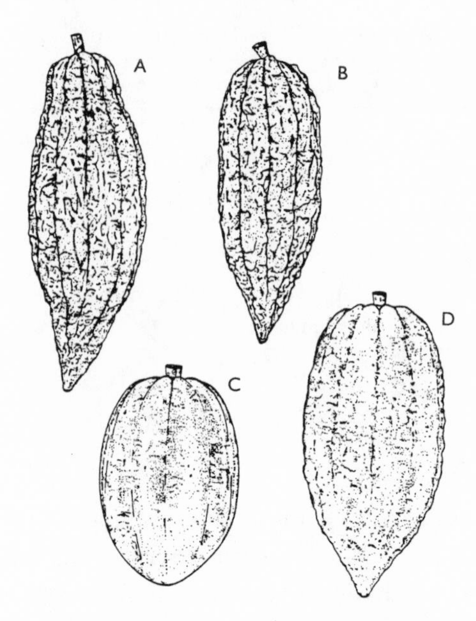

Various major pod shapes in *T. cacao:* (A) *cundeamor*, (B) *criollo*, (C) *amelonado*, and (D) *angoleta*. From Cuatrecasas (1964). Reproduced with permission of the author.

various secondary centers of production on the Atlantic slopes of the mountains, even though little or no cacao is cultivated in these latter areas today. Today the department of Suchitepequez in Guatemala remains an important site of cacao cultivation, as are two other Pacific coast sites. There the trees are hybrids between a native *criollo* and a *forastero* introduced from Costa Rica between 1915 and 1920 (Soria 1966). Small garden plantings of pure *criollo* of the Soconusco type are found occasionally near homes in Guatemala, although, in general, the degree of genetic erosion of the original *criollo* throughout this country approximates 90 percent today (Soria 1959).

What has been strikingly evident with Central American endemic cacao populations is the high degree of local forms occurring spontaneously in the basic *criollo* type found throughout much of this region. Lowland Central America is therefore viewed as a secondary center of genetic diversification in cacao, encouraged by human manipulation, as manifested in the wide range in levels of agronomic vigor and pod morphology within the general *criollo* type, and with considerable differences between many local populations. Types of *criollos*, such as Mexican, Nicaraguan, and Colombian are highly susceptible to disease and insect pests. This may be a result of their ancient, lengthy cultivation by Maya Indians over two thousand years. Such a process cannot be confused with naturally arising varieties because these cultivated races were the products of the manipulation of cacao populations by early peoples, especially in Mesoamerica: the southern half of Mexico, Guatemala, and Belize and eastern Central America as far south as northern El Salvador.

Forasteros, the dominant types of South American cacao, are derived from *T. cacao* subsp. *sphaerocarpum* and include cultivated, semiwild, and wild populations (Wood and Lass 1985). The main kinds of *forastero* cacao are *amelonado*, *común*, West African *amelonado*, *cacao nacional*, Matina (also called Ceylan), and Guiana wild *amelonado*. *Trinitario* cacao descended from a cross between *criollo* and *forastero*, although its features are more predominantly *forastero*. As stated earlier, *trinitario* types do not occur in a wild state, unlike *criollos* and *forasteros*.

The pods of *cacao nacional*, derived from Amazonian *forastero* cacao, closely resembled those of wild cacaos found in the rain forest between the Napo and Pastaza rivers in the Amazonian region of Ecuador (Pound 1938; Soria 1966). This form of cacao, popular today on the international market because of the fine quality of its beans, was most likely in cultivation in Ecuador during the early part of the seventeenth century by Spanish settlers. The greatest expansion of its production took place in the latter half of the nineteenth century. Ecuador at one time was the world's largest producer of commercial cacao. During the first quarter of the twentieth century production of *cacao nacional* decreased drastically because of the decimation of Ecuador's plantations from fungal diseases, to which this variety of cacao is highly susceptible. Today there are no plantations of pure *cacao nacional*, although hybrids with other *forasteros* prevail in Ecuadorian plantations.

Forastero cacao, especially *cacao común*, represents close to 90 percent of the cacao in the expansive plantations of Bahia and Santo Spírito in Brazil. The melon-shaped pods of *cacao común* are phenotypically quite similar to those of Matina cacao in Costa Rica; *amelonado* from Surinam, Costa Rica, the Dominican Republic, and other countries in the Antilles; and Ceylan or Sánchez cacao from Mexico and Guatemala. Thus, although *forastero* cacao is of South American origin, various forms of it have been domesticated and dispersed into Central America and other regions to enhance production in plantations because it is generally easier to grow *forasteros* than *criollos*. The widespread *amelonado* of West Africa is predictably very similar to *común*, since the former was derived from Brazilian cacao through São Tomé and Príncipe. The pod of *cacao común* is much akin to that found in populations of wild cacao in the Obidos region of the Amazon. A cacao cultivar with a more restricted distribution in Brazil is Pará cacao, an Amazonian type bearing highly spherical light green pods, called *calabacillos*, and considered to be a high-yielding variety.

Because the conventional classification of cacao types is based on morphological features and geographical distribution (Enríquez and Soria 1967), genetic and molecular analyses are needed to confirm the degree to which, if any, these many forms of cacao fall into the three agricultural races: *criollo*, *forastero*, and *trinitario*. Even though these three distinct groupings have been in practice for a long time, genetic analysis could reveal a pattern of continual variation, rather than discrete forms, in *T. cacao* at the molecular level.

It is important to realize that horticultural races of cacao most likely bear little affinity to wild species of *Theobroma*, including *T. cacao*. It is difficult to ascertain the degree of human-mediated hybridization of various forms within *T. cacao* that likely occurred in ancient Mesoamerica and the degree to which other species endemic to the region, such as *T. mammosum* and *T. simiarum*, might also be implicated in crossing trials performed thousands of years ago. It is therefore difficult to conclude that what we call *T. cacao* today is really a human-bred amalgamation of many types and species, with no true counterpart existing in the wild.

Cacao Farming Today

Cacao cultivation and production today, the basis for the world's supply of chocolate, is a matter of modest to small-scale farming, chiefly in the humid and wet tropics. The cacao grower faces the challenge of nurturing the trees, in the plantation setting, to yield as many beans as possible, and in the cheapest manner. Under reasonably good soil conditions, a single mature cacao tree may produce hundreds of pods in a year, although thirty or forty pods is more the average. Ten harvest-size pods generally yield one pound of dried cacao beans. With proper local conditions of climate, soil, and care, a single acre of cacao can yield about one ton of beans.

The challenge of obtaining satisfactory yields from cacao plantations has permeated the history of cacao cultivation into modern times. The pervasive uncertainty of cacao harvests was noted not only by the Spaniards in Mexico and Central America in the sixteenth century but later by Humboldt (1884) in Venezuela. He observed that cacao harvests there were variable, affected by unfavorable weather and the large numbers of animals—parrots, monkeys, agoutis, and squirrels—that eat the pods, which are also attacked by various insects. In modern times, pod losses from insects, disease, birds, bats, monkeys, squirrels, and theft are often high.

In Amazonia, where the greatest diversity of spontaneously occurring varieties or races of *T. cacao* is found, researchers have set out at various times to track down new natural varieties as a basis for developing forms with better resistance to disease. One of the most recent cacao collecting programs was the London Cocoa Trade Amazon Project (Allen and Lass 1983). Such efforts recognize that the available cultivated varieties constitute only a small fraction of the total genetic diversity of wild populations of cacao. The object is to harvest a greater share of this gene pool as a foundation for improved selective breeding. Even though the domestication of cacao very likely took place at the northern extreme of its geographical distribution, breeders return to the tree's evolutionary birthplace and center of infraspecific diversity to collect wild forms to selectively cross with domesticated types of cacao. This is done under the perennial veil of mystery that enshrouds our understanding of the basic genetics of the cacao tree. Without a full understanding of cacao genetics, it seems very difficult, if not outright impossible, to even wish for a "super" variety of cacao that will perform to an agronomic optimum. Yet it is not an entirely pointless effort, since it is possible to develop a superior form of cacao by making crosses and hoping for best results, even without fully understanding the genetics. This is analogous to Gregor Mendel's crossing peas in the monastery garden and coming up with hybrid forms long before any knowledge of genes was available. Long before Mendel and the subsequent rise of modern genetics, early peoples undoubtedly propagated the most desirable forms of cacao by vegetative means rather than cross-breeding them, even though they may have initially obtained them by crossing different forms.

Most of our knowledge of cacao production stems from studies in plantations and not in the wild. Much of this agriculturalized natural history focus has revolved around pollination coupled with seed production, pests, and diseases. Yet *T. cacao* has a much longer history as a wild species than it does as a domesticated form of agricultural importance. Any constraints or limitations on the capacity of cacao to produce high yields of cocoa beans in an agronomic or agricultural context, that is, as a domesticated tree cultivated in plantation settings, is going to reflect, in large measure, its natural history in its wild state or condition. The tree's natural history, especially its adaptive heritage as a wild species in tropical rain forest, is the foundation for its performance, or lack thereof, as a domesticated form serving the world's appetite for chocolate. What genetic and phenotypic traits might best serve cacao well as a resident, native species of Amazonian rain forests do not equip domesticated forms of this tree very well as plantation crops. This assertion becomes apparent in exploring the natural history of cacao and studying its pollination biology.

Cacao and Agriculture in Costa Rica

The cultivation of cacao in Central America was eventually transformed from a small-scale, localized practice into a somewhat larger agronomic enterprise. And through this agricultural history, cacao's fate became intimately tied to other events in the development of the region. People and their ambitions, rather than nature, became a prime force in shaping the destiny of cacao. The development of cacao farming and related agriculture in Costa Rica provides a microcosm of the journey cacao has taken throughout much of the humid tropics.

Cacao

In Costa Rica the original areas of cacao cultivation, according to archaeological evidence, were the Nicoya Peninsula on the Pacific, Sixaola on the Atlantic, and the plains of Los Guatusos, immediately south of Lake Nicaragua in the north (Bergmann 1969). Even today, pockets of cacao grow along shaded river banks in dry areas of Nicoya, near Las Cañas. Indeed, there is a river there called El Cacao.

The earliest description of cacao cultivation in Costa Rica comes from Juan Vásquez de Coronado, the original conquistador in Costa Rica. In his sixteenth-century account of the Indian province of Quepos on the Pacific slope of the Cordillera Central, he refers to finding "cacao and textiles, like that of Nicoya" (Fernández 1886). On his fourth voyage to the New World in 1502, Columbus observed Indians paddling huge dugout canoes brimming with giant piles of cacao beans and heading north along the Caribbean coast between Costa Rica and Nicaragua. These shipments of beans, most likely destined for Mexico, were clear evidence of advanced harvesting of cacao beans in Costa Rica by the beginning of the sixteenth century.

In early colonial times, Spanish ships sailed up the San Juan River to the settlement of Grenada on Lake Nicaragua. Unlike the formidable Atlantic Zone, with its dense blanket of rain forest, this region is tropical dry forest. During the dry season, when many of the trees drop their leaves, it was especially easy to penetrate the dry forest and explore the Pacific region of Costa Rica. Vásquez de Coronado established a settlement there at Esparta in the Pacific foothills, from where Spanish colonists forged further inland, across the mountains, to settle in Ujarras and Cachi, where the oldest church in Costa Rica still stands in the picturesque Orosi Valley near Cartago. Although the alluvial lake river bottom of the Orosi Valley held promise of successful farming, the sheltered valley was also ideal for breeding malaria and yellow fever. Settlers moved up the mountainside and away from the threat of disease to a place they called "paradise," known today as the city of Paraíso.

After the Conquest, the descendants of the conquistadors became the owners of cacao *fincas* in the Atlantic lowlands of Costa Rica. Indians were often left to operate the *fincas*, since the owners, who lived in the highlands, did not like the incessant rain and heat of the lowlands, not to mention the snakes, jaguars, deadly fevers, and even fights with Indians. The owners lived chiefly in the more temperate climate of Cartago, the original capital city of Costa Rica, and journeyed down the Río Reventazón Valley to the lowlands once or twice a year to inspect their *fincas*—in time to collect the crop and to plan with their laborers for next season. The journey was perilous at any time, but safest during February and March, when the weather was driest and there was less chance of disease.

The origins and attitudes of Costa Rica's early Spanish settlers help explain, in part, the unique, independent nature of the country in Central America today. The region of the isthmus that would become Costa Rica was not as heavily inundated by the Spanish military as were other regions to the north. Instead, most of the first settlers were farmers. Some evidence suggests that some of these first settlers were Sephardic Jews escaping the Spanish Inquisition. Many Jews resettled in the eastern part of the Mediterranean, but some migrated to the New World, especially to Medellin, in Antioquia, Colombia, and to Costa Rica.

These early settlers typically planted cacao along the banks of rivers, chiefly in the vicinity of the Río Matina in the Atlantic lowlands. The quality of these plantings was very high, and the cacao *fincas*, while small and demanding great care, nevertheless prospered. The eventual establishment of cacao plantations in the Atlantic Zone consequently paved the way for the overall development of agriculture in the humid or wet tropics of Costa Rica.

In 1588, with the destruction of the Spanish Armada by the English, the tide turned for the New World colonial settlements. The symbol and the reality of the Spanish Empire's invincibility were gone, and the age of English pirates had arrived. English seamen raided Costa Rica, which was still part of the Spanish Empire, and absconded with cacao, by this time a prized commodity in Europe. The chief buccaneer, Henry Morgan, came from the Caribbean coast to sack towns as far inland as Turrialba, which was not very far from where cacao harvests were stored in Cartago. Frequent raiding parties discouraged the settlers' interest in cacao farming, and cacao production remained doggedly low for many years.

Coffee

Still, cacao was Costa Rica's major export crop until the nineteenth century, when it was largely overshadowed by coffee. Coffee was brought into Costa Rica from Jamaica or Cuba in 1808 and planted in small *fincas* in the highlands by those who could afford to cultivate the new crop, called *café* by the Costa Ricans. Many of the small *fincas* were located on the leeward or Pacific slopes of the Cordillera Central and in mountains northwest of this region.

Coffee is originally from the highlands of Ethiopia, where the bush is endemic to forests between 3,000 and 4,000 feet in elevation. The use of coffee as a beverage started with Arabs and Ethiopians, who brewed the leaves from the bush. Later, it was discovered that the dried beans, when ground up, created a more potent beverage. And the beans were easier to handle and ship great distances.

Coffee drinking was established in Constantinople in the fifteenth century and was popular in England by the sixteenth century. With the rising popularity of coffee in Europe, the Dutch planted coffee bushes from Ethiopia in Indonesia, chiefly in Java, as part of the Dutch East India Company. Because Indonesia was the world's chief source of coffee before the mid-1800s, the beverage became known worldwide as "java."

In an effort to increase yields of coffee beans, Dutch agronomists shipped new plants from Ethiopia's upland forests to Java, in the process importing the dreaded coffee rust, a disease produced by the fungus *Hemileia vastatrix*. The rust soon eradicated much of the coffee plantations in the Dutch Indies early in the nineteenth century. That, together with increasing market demands for coffee in Europe and North America, helped the fledgling coffee industry take hold in Costa Rica.

On Christmas Day in 1843, William Le Lacheur, a trader returning to England from British Columbia, stopped off at Puntarenas on the Pacific coast of Costa Rica to obtain ballast for his ship. Venturing inland from the coast, Le Lacheur encountered farmers with a small coffee crop they were unable to sell. Transportation to the coast was their biggest obstacle. The sixty-five-mile journey across the rugged mountains to Puntarenas depended on oxcarts and mules. Le Lacheur bought their entire crop on credit, offering to pay the farmers on his return from England. After a three-month voyage around the Straits of Magellan, he returned, paid off his creditors in Costa Rica, and sold off various goods brought back from England. With the profits from these sales, Le Lacheur was able to purchase larger harvests of coffee beans, which helped prompt the expansion of coffee agriculture in Costa Rica.

By the middle of the nineteenth century, coffee commerce in Costa Rica burgeoned with the creation of an oligarchy of coffee growers. New towns and roads for transporting the harvests arose across the rugged hinterlands. Many bridges were built across steep mountain streams and rivers. Steady trade of Costa Rican coffee with the outside world had begun. For the first time, iron stoves and glass window panes made their way into Costa Rica, greatly changing people's lives. Government decrees were set forth requiring each municipality in the uplands to complete the planting of coffee bushes in all predesignated acreage within twenty to thirty years. Each municipality, under official edict, was required to plant ten *manzanas* (seventeen acres), in the highlands up to 1,500 meters elevation, above which coffee bushes do not grow well. This intense effort to expand coffee production in Costa Rica left its mark—70 percent of the coffee *fincas* in Costa Rica today were planted during this short period.

One of the many budding coffee barons of the time, Julio Sánchez Lepis, developed a lucrative business of transporting coffee beans to Puntarenas. With the profits from their businesses, Sánchez and others eventually bought many coffee *fincas* in Alajuela in the Central Valley, realizing that plans to build a railroad to the Caribbean coast would boost the coffee industry in Costa Rica even more. Today, many coffee *fincas* around the towns of Alajuela and Heredia still have old iron gates with the initials "J. S. L." on them. A daughter of Sánchez married an Arias, and their son, Oscar Arias Sánchez, grandson of the coffee baron, be-

Overview of a Costa Rican coffee plantation in the highlands near the city of Turrialba, Cartago Province, Costa Rica.

came president of Costa Rica in 1986 and received a Nobel Peace Prize the following year.

In 1871, General Tomás Guardia, the president-dictator of Costa Rica, dreamed of making the transportation of the growing coffee harvests more efficient and profitable. There were visions of increased revenues from a government-owned railroad that would transport the beans from the Meseta Central to the Caribbean port city of Limón, greatly shortening the arduous sea voyage from Puntarenas to North America and Europe. The rugged terrain of the mountains between the Meseta Central and the Caribbean coastal region challenged Guardia's ambition. But there was great impetus for such a plan: Banks and industrial concerns were being formed to handle the increasing exportation of coffee beans, and foreign capital was pouring into the country for the first time.

Much of the land between the Meseta Central and Limón was still clothed in virgin tropical rain forest. It proved to be a formidable task to open the region for a railroad. Local folklore has it that for each crosstie on the twenty-mile stretch between Limón and Matina, one worker died from malaria, yellow fever, dysentery, or a machete fight. Aside from the cacao farmers in this region, there was not much of a Costa Rican population in the area at the time to assist with

Ripening *arabica* coffee beans in the central highlands of Costa Rica.

the railroad, and most of the rail workers were brought from Barbados and Jamaica, in addition to the Chinese and Italians brought in from the Caribbean. The Jamaicans were only one generation removed from slavery, having been granted new status as British subjects in 1834. Their descendants live in Limón today, as do many descendants of the Chinese. Up and down the Caribbean coast of Central America, from Belize to Panama, there are many communities of descendants of Jamaican and other West Indian blacks—testimonials to an era in which the economic backbone of Central America, its agriculture, was being developed.

Imagine what it must have been like for the workers to carve out a railroad while confronted with the endless rain, mudslides, fallen bridges, humidity, and oppressive heat of the Caribbean plains. Consider, too, another drama that surely unfolded before them, one that might have made the misery easier to handle—the natural beauty of the untouched rain forests. The only openings in the rain forests were generally natural clearings and small cacao *fincas* scattered here and there, carved out a century or two before. Did these men hear the night call of the jaguar and the crash-diving of monkeys from the tops of lianas along the trail they blazed to lay the track? What manner of butterfly, *Morpho*, *Caligo*, *Prepona*, or others, came to the feed at the men's own dung at the edges of their work camps? How majestic it must have been to stand in the shadows of the giant trees and see the *escalera de mono* (monkey ladder) vines for the first time. And how terrible it must have been to become acquainted with the sting of the *bala* ant, hornets whose nests they accidentally crushed, and the deadly bite of the fer-de-lance.

Costa Rica was destined to have a railway, and eventually it did, changing the country's economic destiny. The railway ushered Costa Rica into a new era of agricultural commerce with the outside world when it opened in 1890. It provided a new point of departure and entry for harvested crops and goods on the Caribbean, with greater direct access to Europe and well-established U.S. seaports like Boston and New York.

The newly finished railway reached west from Limón for 190 kilometers to Alajuela, providing a vital link for people and commerce to and from the bustling Meseta Central to the desolate, isolated Caribbean floodplain. Everyone would use the railway: people en route to the local medical clinic, children and teachers going to school, farmers with their livestock, tourists. Most importantly, it became the economic link to the outside world, providing a means of transporting coffee, cacao, and other crops to Limón, where these exports, vital to the economy of a developing Costa Rica, could be shipped across the seas.

Throughout much of the grueling railway project, bananas, a developing agricultural crop of the Atlantic lowlands, became another incentive for building the railway. Bananas would be shipped by rail from plantations in the interior to Limón and loaded onto schooners bound for New York, Boston, Europe, and other ports on the global trade routes. Costa Rica's expansive Atlantic zone, especially the Limón area, became the land of bananas largely by accident. Early merchants and entrepreneurs, who planted bananas along the railroad still under construction, used the profits to finance the completion of the project.

Three centuries earlier, on his fourth voyage to the New World, Christopher Columbus landed close to Limón at the Indian village of Cariari. Columbus thought that he had reached the fabled Siam, where he would plunder the gold mines he had read about in the writings of Marco Polo. Impressed by Indians wearing a few gold trinkets, Columbus called the land Costa Rica (rich coast), only to discover that there was little or no gold to be found. But the region did later become the source of Central America's other gold—bananas.

The story of the origin, domestication, spread, and arrival of bananas in the New World tropics is a complex one, and only a bare outline is offered here (see Simmonds 1959 for a full account). Cultivated, edible bananas have their origins in the hybridization of two wild species of the Indo-Malayan region, *Musa acuminata* and *M. balbisiana*. During the first millennium B.C., hybrid polyploid bananas were introduced into East Africa. The first records of their cultivation come from India, in 600–500 B.C. Bananas were chronicled by the Chinese in A.D. 200 and introduced to the Mediterranean in A.D. 650 and Polynesia in A.D. 1000. During the early fifteenth century bananas were introduced into the Canary Islands from West Africa by the Portuguese. And in 1516, following the Spanish Conquest, bananas were introduced from the Canary Islands into the New World, to the Caribbean island of Santo Domingo.

The first varieties introduced in the West Indies were "Silk Fig" and "French Plantain," which, according to Linnaeus, were *Musa paradisiaca* and *M. sapientum*, the latter also called "Muse of Wisdom." Other well-known varieties such as Gros Michel and Dwarf Cavendish were introduced into the West Indies in the nineteenth century. Another species of banana, *M. textilis*, is the source of manila hemp. Plantations near the Costa Rican towns of Manila and Bataan, near Limón, grew *M. textilis* to provide hemp rope for battleships during World War II.

The arrival and establishment of bananas in the West Indies and eventually

A banana plant with ripening bananas, along the border of a Costa Rican cacao plantation.

into mainland Central America, especially Honduras and Costa Rica, led to the banana export trades of the nineteenth century. Schooners carried the fruit to U.S. ports during spring and fall, when there was the least chance of spoilage. Between 1800 and 1849, bananas were in cultivation on a limited scale in several Caribbean islands. By 1850, bananas were being cultivated in Panama, where the first Central American railway was being built. The first regular shipment of bananas to the United States is accredited to the brothers Frank, who exported the fruit from Panama (still part of Colombia at the time) to New York in 1866.

With the establishment of the Boston Fruit Company in 1885, the banana trade soared. In Jamaica, the center of the fledgling venture, small farmers were encouraged to expand their plantings of the exotic fruit. The Boston Fruit Company also expanded plantings of bananas to Santo Domingo and Cuba. By the mid-to-late 1890s, ten million stems of the fruit were being imported to the United States each year from Jamaica, Cuba, and Santo Domingo. Steamers, eventually outfitted with refrigerated storage holds, soon replaced schooners.

With the completion of the Costa Rican Railway in 1890, Central American trade, chiefly to gulf ports such as New Orleans, burgeoned. Nine years later, the United Fruit Company, the giant banana company of Central America, was founded in New Jersey by the amalgamation of the Boston Fruit Company and the Costa Rican holdings of Minor Keith, who, in 1874, was the first to plant bananas in Costa Rica during the railway project. He later became the "banana king" of Costa Rica, establishing himself as a principal in United Fruit.

The Atlantic zone proved to be an ideal environment for the crop. Many independent banana growers appeared in Costa Rica, pushing back old plantations of cacao to make way for giant plantations of bananas and peeling away more and more of the Atlantic zone's rich blanket of tropical rain forests by fire, axe, and machete. By 1916, twenty-six years after the completion of the railway, bananas were a huge economic success in Costa Rica, which experienced a tremendous expansion of plantations and transportation. In 1916 alone, about sixteen million stems of bananas were exported from Limón, which came to be synonymous with the popular fruit. From these humble beginnings in the Atlantic zone of Costa Rica, the banana industry spread quickly into Nicaragua, Honduras, Guatemala, Panama, and Colombia.

When the rain forests of the Atlantic zone were replaced with large-scale banana plantings, the lives of many creatures were disrupted. The sheer abundance of the banana trees in these big plantations, together with the close relatedness of *Musa* and *Heliconia* (wild plantain), likely facilitated an ecological shift in caterpillar food plants of certain butterflies from natural hosts to the exotic bananas. The owl butterfly, *Caligo*, had dwelled in the primeval forests for thousands of years before the arrival of banana trees. This butterfly, a host of other insects, and fungal disease organisms were all compelled to exploit this new, abundant food supply. And disease organisms eventually spelled disaster for the banana industry.

In 1916, Panama disease (*mal de Panamá*, caused by the fungus *Fusarium cabense*) was discovered in the Limón area plantations. It attacked the popular variety, Gros Michel. By 1926 the root disease was so rampant that exports were down to about one million stems a year, compared with five million to ten million in previous years. With the ravages of Panama disease in the Atlantic zone, banana operations in Costa Rica expanded to the Pacific coastal region.

The Banana-Cacao Pendulum

Diseased banana farms in the Atlantic zone were then replanted with cacao, and the best varieties of cacao were brought into Costa Rica for this purpose. Large-scale cacao bean processing facilities, driers, and fermentaries were built at Limón and at Almarante, just across the border in Panama. By 1948, more than 70,000 hectares (172,000 acres) had been planted in cacao. For a short while, Costa Rica became the world's tenth largest producer of cacao beans, even though its total production represented only 1 or 2 percent of the world's total. At one point during the cacao fever, Costa Rica exported a record 16,000 metric tons.

But cacao's success would only be short-lived. As cacao production in Costa Rica soared in the first quarter of the twentieth century, another fungal disease was brewing up a different ending to the script: *Phytophthora* devastated cacao pods, rotting 50 to 80 percent of each harvest. Even when cacao trees are sprayed to combat the black pod disease, losses can still be high.

Interest in growing cacao declined even further in Costa Rica after World War II. Disease, pests, repeated thefts of crops, and sagging bean prices compelled the United Fruit Company, which owned the largest plantations of cacao in the region, to turn over much of the land to former employees as their work contracts terminated. Much land was also turned over to the Costa Rican government. Processing operations at the big bean dryer and fermentary in Limón were shut down, and United Fruit holdings near the Pacific Coast were converted to African oil palm plantations—a source of oil for margarine.

At the same time, the Costa Rican government enacted a 13 percent export tax on cacao beans as a means of generating revenue to improve roads and build schools and hospitals in the Atlantic zone. In addition, Limón Province, the center of the country's cacao industry, decided to impose its own 3 percent tax on exports. For many, it was time to bail out of the cacao business.

In the 1950s and 1960s, with cacao prices at new all-time lows, and the stagnation of the cacao industry in the lowlands, banana fever took hold once again in Costa Rica. New varieties of bananas resistant to Panama disease, the Cavendish and Valery, reinvigorated the banana industry. The Standard Fruit Company came into Costa Rica from Honduras, developed large plantations of the Cavendish, and began boxing the fruit for shipping. Limón commerce and railroad activity picked up.

In the following decade, cacao made a partial comeback as Soviet-bloc nations began importing cacao beans. This surge of interest brought with it investments on the part of producers to improve cacao plantations and plant new trees. By this time, new hybrid varieties of cacao, with greater promise for better disease resistance, high productivity, and precocity had been developed. In 1979, Gill & Duffus, a leading commodities group for the cacao industry, predicted a record ten-million-pound crop for Costa Rica, loosening local bank credit for cacao *finceros* in the Atlantic zone. But success again proved to be short-lived. Yet another fungus disease, *Moniliopthora,* brought into the Limón area accidentally from South America, began attacking pods and initiated a sharp decline in production, which spelled disaster for Costa Rica's cacao industry.

A tremendous decline in the world market price for cacao beans in the mid- and late 1980s again added to cacao's decline in Costa Rica. From an impressive all-time high of $1.90 per pound on the New York spot market in 1977, cacao bean prices dropped to under ninety cents per pound by 1987. The United States, the world's largest importer of cacao beans (205,000 metric tons in 1986), lowered its quota by about 12 percent as market traders waited for the price to decline further.

Cacao production is a matter of hard economics, with wild swings in the availability of cacao beans, market prices, and consumption. For example, the U.S. Department of Agriculture forecasted that the global 1986–87 cacao bean harvest would exceed 1.9 million metric tons, while consumption was expected to be about 1.8 million tons, leaving an estimated surplus of 100,000 tons. Americans alone typically consume about eleven pounds of chocolate per person each year. Added to previous stockpiles, the total supply of beans would be about 700,000 tons, according to Gill & Duffus. But forecasts of production levels can be devastated by bad weather and other environmental factors. Drought, disease, and political turmoil periodically decimate the annual harvest, causing prices to go way up. Overproduction obviously pushes down the price.

Events in the early 1990s portend a possible shift towards increased cacao production in Costa Rica and elsewhere, even though current world market prices for cacao beans (1990–91) are again low, about $2,000 per metric ton, chiefly on account of overproduction in the Far East. Cacao production in Malaysia, for example, rose from 26,000 metric tons in 1978 and 1979 to more than 225,000 metric tons in 1988. But it has been predicted that the price of cacao beans will increase over the next decade, to more than $4,500 per metric ton (Hunter 1990). Also, by the fourth quarter of 1991, the global cacao bean surplus was in sharp decline, prompting an increase in world interest to produce more cacao. An earthquake in the Atlantic zone of Costa Rica in April 1991 destroyed extensive banana operations, prompting growers to open new banana plantations in the northeastern district of Sarapiquí. Such conditions eventually may also encourage renewed planting of cacao throughout the Atlantic zone. In the long term, land areas degraded with banana fields ought to be replaced with native tree species in well-planned reforestation programs, possibly integrating cacao as part of such an effort.

Rubber

Pará rubber is native to the Amazon rain forest in Brazil, which at the time was the leading supplier of the crop in the free world. But by 1940–41 the Japanese had advanced to the major rubber-producing islands of Java and Sumatra, and even to Singapore, thereby choking off much of the supply of rubber to the West. There was great incentive to develop rubber plantations in the American tropics as a result. Eventually, U.S. research during the early 1940s led to the development of a synthetic rubber, an innovation that greatly diminished the demand for natural rubber, *Hevea brasiliensis*.

A French botanist named Fusee Aublet may have named the genus *Hevea*, derived from the Carib Indian word *heve*, in reference to a different tree, *Castilla* (Moraceae). Pará rubber is still called *heve* in Ecuador and Peru. *Hevea* is in the spurge family, the Euphorbiaceae, which includes the familiar poinsettia. It is a crop of the humid or wet tropics and native to the American tropics. Aublet and various other botanists collected extensively and recognized as many as twenty-four species of *Hevea*. According to Richard Schultes, who worked in the Amazon for many years before World War II, the genus contains only nine or ten valid species, of which three—*H. guianensis*, *H. benthamiana*, and *H. brasiliensis*—yield commercially valuable latex and have been harvested in the wild. More recent estimates indicate about a dozen species of *Hevea*.

H. brasiliensis produces the highest yields of the prized latex by far. In Brazil, the latex of *H. brasiliensis* is called *borrachea* (bottle), but more commonly *serigna* (syringe)—names derived from ancient Indian practices in Amazonia, where small bottles and syringes were fashioned from the latex and filled with water to be used as toys (squirt guns) at fiestas. Portuguese conquistadors who came to Brazil in the sixteenth century adopted these Indian names for the tree and its latex. Columbus, on his second voyage to the New World, wrote of seeing Indians in Haiti playing with rubber balls made from the latex *Castilla* tree.

In spite of its origins in the American tropics, Malaysia became the world's leading producer of natural rubber. It was the British who brought rubber trees into Malaysia from Brazilian Amazonia in the early 1900s. Harry Wickham, the British botanist and agronomist, who was later knighted, transported 20,000 rubber tree seeds from Amazonia to the Royal Botanic Gardens, Kew, Great Britain, where the seeds were successfully germinated in "glass houses." The first seedlings were then shipped to Ceylon (now Sri Lanka) and from there to Singapore and Kuala Lumpur in Malaysia, where only eleven rubber seedlings survived. Under the guidance of Henry Ridley, the farming and processing of rubber advanced to a very successful commercial stage at what became the Rubber Research Institute of Malaysia. Ridley trained many of the local people in the culturing and tapping of rubber trees, making the operation the leading high-quality producer of rubber in the world, which it still is today.

Along Costa Rica's Atlantic slopes and floodplains, occasional plantations of Pará rubber are snuggled in tightly between wedges of rain forest. In Sarapiquí, at Finca El Uno, cacao trees grow beneath tall, stately rubber trees planted long ago by Bob Hunter. Hunter also planted rubber trees on the nearby land of José Heinrich von Storen in Sarapiquí. These and the experimental planting in Turrialba are the only stands of *Hevea* left in all of Costa Rica today. Cacao and Pará rubber trees fare well together, especially since their pollinators belong to the same family of Diptera. The dense, matted leaf litter established by the *Hevea* and cacao association is suitable for the breeding cycle of these insects, as are the tank bromeliads and other water-holding epiphytes often found in the branches of the rubber trees.

CATIE and Cacao Research

Successful production of cacao in Costa Rica and other parts of the tropics has been fostered by basic agronomic research. Costa Rica has played a key role in

A Pará rubber plantation in Sarapiquí. The rubber tree in the foreground is being tapped to collect the prized latex.

attempts to improve cacao cultivation through establishment of internationally supported teaching and research programs within the country. Since 1950 research on cacao and other crops has been linked at various times to the Centro Agronómico Tropical de Investigaciones y Enzeñanza, or CATIE, an agricultural-forestry institution in Turrialba.

CATIE traces its roots to 1939–40 and a cooperative venture between the United States and Costa Rican governments to introduce cultivation of rubber into Costa Rica. President Franklin D. Roosevelt's newly hatched Good Neighbor Policy advocated a U.S. Department of Agriculture research station in Latin America. With the world still dependent upon natural rubber, and the threat of war on the horizon, the initial focus was expanding and improving the production of rubber. Thus, the Regional Rubber Institute was established in the Turrialba Valley. It became known as Beltsville South—a reference to the U.S. Department of Agriculture's research facilities in Beltsville, Maryland. The first director of this station was Dr. Ernest P. Imle. Imle had come to the rubber station in Turrialba in 1939 from the Boyce Thompson Institute in Yonkers, New York, and many years later became one of the early scientific advisors to the American Cocoa Research Institute. Imle played a major role in the formation of the Instituto Interamericano de Ciencias Agrícolas (IICA), an agricultural research university, on land next to the rubber station in the Turrialba Valley recently expropriated from Germans.

The Rubber Institute would serve as a research and teaching center for agriculture not only in Costa Rica but throughout Latin America. The original team of people at the Rubber Institute formed an alliance with Charles Averry, the head of the Northern Railway in Costa Rica. In 1940–41, the Northern Railway built the landmark headquarters building and faculty houses on the newly formed institute grounds. Both the white stucco main building and the stucco houses, with their tile roofs, are still present today. Averry and his crew created the small lake on the institute's grounds by pushing back a swamp, the remnants of which are still found today around the lake's perimeter. Ralph Allee, the institute's second director, assumed office in 1943, advancing the institute's mission of providing training and research in Latin America's principal crops: cacao, rubber, bananas, and coffee. The envisioned Beltsville South had really come into fruition. In 1956 the old rubber station combined with IICA. In the early 1960s, the director of IICA, a Colombian named Armando Samper, moved the IICA offices to San José and renamed the institute's Turrialba facility CATIE. IICA operated in all Latin American countries of the Caribbean region, not just Costa Rica.

Once expropriated land was acquired, the owners of the nearby King Ranch donated funds for a laboratory to study tropical cattle farming. The project started in 1941 but was interrupted on December 7 at the time of the Japanese attack on Pearl Harbor due to a sudden shortage of steel and building equipment needed for the war. The war effort boosted research on rubber, bringing to the institute in Turrialba a number of prominent U.S. scientists to study ways to increase production of the needed wartime commodity.

Post-war euphoria swept across the United States, and the country was determined to help build the world, exporting to places like Latin America its skills and talents in agriculture. Part of this effort was channeled through the Institute of Interamerican Affairs (IIAA), coordinated by Nelson Rockefeller, and including an agriculture and extension program linking U.S. technical assistance with the Costa Rican agricultural community. IIAA had been proposed in President Harry Truman's January 20, 1949, inaugural address as part of the Four Point Plan. Under this plan, approved by the U.S. Congress in May 1950, the United States declared its commitment to provide technical assistance to developing countries. The fourth point of this plan called for the United States to expand technical assistance programs throughout the world. Agronomists such as Bob Hunter were hired under this program to assist Latin American nations. After finishing his Ph.D. at Michigan State University, Hunter headed for Costa Rica, where he worked for IIAA from 1951–55. The primary mission of this institute was to research ways to combat and control leaf blight of Pará rubber trees.

In Costa Rica, STICA (Servicio Técnico Interamericano de Cooperación Agrícola) was eventually created as a hybrid, liaison organization between the Four Point Plan and the Costa Rican Ministry of Agriculture. Bob Hunter became director in 1955. The program included as its major mission the development and expansion of an agricultural extension service. In the United States, such an effort had been a key to successful agricultural development, particularly when it was brought together with land-grant colleges to provide education and research, functions that go hand-in-glove with extension work.

Hunter ran STICA through the Foreign Operations Administration (FOA) as a U.S. State Department employee in charge of a bi-national agriculture and extension program. He wanted to infuse more of a research component into the program to complement and strengthen its educational and extension components. As director of the program in 1955 and 1956, he headed the building of a soils laboratory and geology building at the University of Costa Rica campus in San Pedro. He developed several centers for the propagation of cacao in Costa Rica, including ones at La Lola, Guapiles, San Carlos, and Puerto Viejo de Sarapiquí. One of these cacao propagation centers, Finca Experimental La Lola, was the principal research site for my later studies on cacao pollination. The La Lola Farm, near Siquirres in the Limón Province, was donated to CATIE by the United Fruit Company in the 1960s with financial assistance from the American Cocoa Research Institute (ACRI). Originally a banana farm, La Lola became an experimental cacao plantation with a focus on the propagation of cacao.

In 1961, Hunter left IICA, shortly before it became CATIE, and went to work for two years for the Central American Bank for Economic Integration in San José. During this period, he was a chairman of the executive committee of tropical biology, the forerunner organizing group that eventually gave rise in 1963 to the Organization for Tropical Studies (OTS) and the Associated Colleges of the Midwest Costa Rican Field Studies Program (ACM). Hunter became ACM's first field director in 1963 and stayed in that position until 1974, when he resigned to devote his full time to his farm operations in Sarapiquí.

Over the years, CATIE has become an international resource center for tropical agriculture, exporting information and plants to different parts of the world. As its two main sources of funds, CATIE receives millions of dollars from the United States each year for its many programs under the Agency for International Development and equally impressive monetary support from European

Dr. J. Robert ("Bob," "don Roberto") Hunter, in the doorway to the office of the La Tirimbina Farm operation in 1978. He is wearing a hat bearing the Hershey Chocolate Company logo. At the time of this picture, Hershey was a business partner of La Tirimbina, in particular the La Tigra cacao farm, under a partnership called Huntrosa.

countries. CATIE has been an international symbol of cooperative goodwill, linking Latin America to other parts of the world through its scientific research.

Research on New Varieties

Agricultural research and teaching centers such as CATIE have been important resources for the development of many new varieties or agricultural races of tropi-

The main buildings of the La Lola Experimental Farm as they appeared in 1981. The covered walkway between the station manager's house (barely visible to the left) and the main station shields personnel and visiting scientists from the often torrential rains sweeping into this area from the Caribbean Sea. Various animals, including bats, rats, and opossums, nest in the space between the roof and deteriorating ceiling panels of the upper level in the main station. These otherwise sturdy wooden buildings were originally built by the United Fruit Company for its banana plantation operation more than sixty years ago.

cal crops, including cacao. These centers often maintain living collections, called "clonal gardens," of the many varieties of cacao that have been produced by selective breeding and vegetative propagation of various forms of cacao. The aim of such programs is to produce varieties that perform better than existing ones in terms of yields of cocoa beans.

Much effort in improving cacao production has focused upon the selection of varieties. In 1916, when the United Fruit Company in Costa Rica switched from growing disease-plagued bananas to cacao, it began to develop a series of cacao varieties derived from vegetatively propagated trees of the best cacao from the Limón area, which were all derived from Central American types of *criollo* cacao. The United Fruit Company also imported cacao seeds of the best selections from Trinidad and Ecuador to complement the original varieties with *trinitario*, South

American *criollo*, and *forastero* cacao. Each variety arising from this effort was given a United Fruit designation, followed by a number. Thus, today, familiar cacao varieties in the American tropics include UF-29, UF-221, UF-613, and so forth. The aim of the selection program was to obtain not only the most productive varieties, but also those bearing the largest pods and large seeds.

Yield in cacao is defined by two values: seed index is the average dry weight per seed; pod value is calculated by dividing the number of mature (harvestable) pods by the weight of dry cacao seeds obtained from these pods (Wood and Lass 1985). The pod value and seed index were developed in Trinidad more than half a century ago to provide a reliable means of scoring or ranking varieties (Pound 1933). The lower the pod value, the higher the yield of the seeds. Pod values are computed from individual trees and varieties and define the minimal limits in selecting trees for cultivation to achieve an acceptable yield of cacao beans. The limits of selection in *trinitario* varieties, according to Pound, turned out to be production of 50 to 100 pods per year, with a pod value of up to 7.5 and a seed index of 1.5 to 1.8 grams. In most *forastero* types, the limits were a production level of 100 to 200 pods per year, a pod value of 10 to 12, and a seed index from 1.0 to 1.2 grams.

In the two decades following Pound's work, the Trinidad technique was used in defining the pod values and seed indices for different varieties grown elsewhere. Based on these criteria, many varieties have since been selected for enhanced crop yields, with large clonal gardens of high-yielding varieties now available in Trinidad, Ecuador, Brazil, Costa Rica, Venezuela, Mexico, Colombia, and, more recently, Puerto Rico. The vegetative propagation of these materials, rather than using seed from difficult-to-predict sexual crosses, is the most effective means of selecting the variety or clone best suited for a particular ecological situation. A cacao grower can observe what varieties are doing the best in a particular region and local conditions and vegetatively propagate those varieties to possibly increase yields. Much of the successful cacao production of Africa today is the result of vegetative selections taken from the first-generation progenies of Amazon-derived varieties introduced from Trinidad to Ghana in 1944 (Roberts 1955).

Vegetative propagation, a nonsexual means of selecting and producing varieties, was accomplished by making rooted cuttings, or bud-grafts. But rooted cuttings did not work well since the cuttings do not have a tap root and topple over easily. This deficiency led to programs in which so-called hybrid seed was produced by sexually crossing two parental stocks. For this, cacao flowers are hand-pollinated, a technique developed by Pound in the 1930s. This tedious, labor-intensive technique entails placing a known type of pollen onto the stigma of a flower of a different cultivar using a splinter of wood or fine tweezers. In some instances in Pound's studies, the resulting F1 offspring exhibited a higher yield of seeds when they matured into flowering trees three to five years after planting. Such "hybrids" became very popular for a while with growers since these trees possessed a tap root and a chupon-pattern (basal or trunk shoot) of growth they liked (Hunter 1990).

But the problem with such hybrid seed is that the genetic lineage of the parental lines is generally unknown or poorly known. Because of this, many crosses give unpredictable results in terms of the desired increased yields. Real and consistent improvement of cacao yields through the use of hybrid seed suffered from a lack of controlled field trials needed to confirm assertions that certain varieties gave superior results over others. Part of the reason for this has to do with the long life cycle and generation time of the cacao tree, which takes five years to flower. Other kinds of crops having one or two generations per year, such as corn, wheat, and rice, have responded very well to the development of consistently good hybrid seed. In such food crops, it is much easier to make many kinds of controlled crosses and quickly assess the results. This is not the case with cacao.

True hybrids, in the Mendelian sense, result when two pure lines are crossed and produce phenotypically recognizable heterozygote offspring that exhibit hybrid vigor or other desirable traits. This is not the situation with cacao breeding and selection programs, since it has not been possible to identify truly pure lines. Studies currently under way in the Cocoa Research Unit at the University of the West Indies in Trinidad, in which gene markers are known and can be detected with electrophoresis, offer one possible means for identifying different genetic lineages within previously designated varieties or clones.

The largest and most important cacao-breeding program in the Western Hemisphere was undertaken by the Ministry of Agriculture in Trinidad under the direction of W. E. Freeman, beginning in the 1930s and lasting about fifty years (Hunter 1990). The program began with crosses of only four varieties: IMC-67, SCA-6, Pound-18, and ICS-1. In spite of an initial high incidence of genetic mixing from the heterozygous seed in these crosses, eventual success came from a series of field trials in which only the most promising offspring were retained for further selections. Selections were assessed by their pod values.

Other selection programs came into being to enhance cacao's resistance to diseases, but these attempts have not been consistently successful, largely because of a lack of understanding of the basic modes of inheritance of the genes underlying any such resistance. Typically, when different clones are crossed, the progeny might be initially resistant to a disease, but the resistance is soon lost. Until the genetics of cacao is fully understood, the use of sexually produced new varieties remains a black box charged with highly variable results.

A major effort, with mixed success, to select for and breed varieties resistant to fungal diseases took place in Ecuador in the 1950s. Selections were made from local varieties and ICS (Imperial College Selections) varieties brought from Trinidad and from material collected by F. J. Pound in the 1930s from the Upper Amazon River. The Amazonian material included Pound-12 and the Scavina, or SCA, varieties, all under the headings of EET varieties (Estación Experimental Tropical, at Pichilingue, Ecuador).

The prime impetus for Ecuador's cacao selection program was to combat *Moniliopthora roreri*, a fungus that causes cacao pods to turn black and rot (Chalmers 1972). Earlier, before World War II, the Imperial College of Agriculture in Trinidad developed its ICS clones to combat another fungal disease threatening to wipe out the country's cacao plantations, witches-broom (*Crinipellis perniciosus*), which causes the tender terminal-tip growth leaves and shoots to turn black and shrivel into a tassel. The selections were based on both local material and Pound's Amazonian material. This program resulted in a series known as the TSH (Trinidad Special Hybrid) varieties. Some of these TSH varieties possess a yield potential of more than two tons of cacao per hectare per year and a very high pod value of 8 or 9 (compared to an average pod index range of 4–6).

Today, many different varieties of cacao are recognized, products of breeding programs. The basis or criteria for selecting varieties has included seed size, pod size, number of beans in a pod, resistance to various fungal diseases, rate of seedling growth, and other traits related to agricultural performance.

Hunter (1990) points out that an absence of good records on the selection programs in various countries has added considerable confusion to understanding the origins of the various cultivars used extensively today. Additional confusion has arisen because each country with a selection or breeding program devises its own designations for varieties, even if they are imported. As a result, the same varieties might have different names in different cacao-producing countries where selections have been made even though they are the same selections. Mexico produced R clones, Guatemala GA varieties, Colombia CA and APA varieties, and so forth.

Genetics

Much work on cacao has focused upon inherited characteristics of crop yield components such as number of pods per tree and the dry or wet weight of the individual seeds (e.g., Pound 1932a, 1932b; Bartley 1964, 1968). Attempts have been made over the years to identify the degree of "heritability" of these various components of crop yield in cacao, with mixed and sometimes confusing results. In some of these studies, certain components, such as bean number or size, appeared to be good predictors of yield, while in other studies, other components gave the best results (J. V. Soria 1970, 1978). Attempts at making controlled crosses between distinctive varieties examined the combined effect of the genetic constitution of both parental lines as they affected certain components of yield. For example, in one study the Amazon-derived varieties Pound-7, Pound-12, and IMC-67 exhibited high combining ability in different crosses, while SCA-6 and other *trinitario* clones had medium to low combining abilities for yield (Soria et al. 1974). Such studies revealed that yield in cacao varied greatly between different crosses of varieties and that it was often difficult to accurately predict changes in yield in crosses.

There is still a lack of good or superior stocks (varieties), and little to no investment in genetic research and selection of cacao clones, even though such an effort is needed. A selection program could be developed to take advantage of the cacao "gene banks" of the U.S. Department of Agriculture in Mayaguez, Puerto Rico, and Miami, Florida, as well as the International Cocoa Gene Bank in Trinidad, the cacao collections at CATIE in Costa Rica, and those at Commissão Executiva do Plano da Lavoura Cacauerira (CEPLAC) in Belém, Brazil. These collections, which include recent and new material from the Amazon gathered by John Allen under the London Cocoa Trade Amazon Project in the 1980s, represent virtually all of the variants occurring in *Theobroma cacao* and provide a foundation on which to conduct the needed research on cacao genetics.

Cacao research facilities have tended to focus on devising morphological "descriptors" for classifying different varieties. The significance of this work is that it attempts to develop accurate and reliable descriptions of cacao varieties based chiefly upon the morphological features of the flowers, pods, and other structures. But there is little understanding of the basic genetics underlying any morphologic traits. There is also a lack of reliable data on the relative field performance and ecological requisites for cacao varieties. All of this means that today's cacao growers rely upon vegetatively propagating those varieties which they observe do the best under the conditions on their individual *fincas*.

Such homespun cacao selection programs must be very similar to what aboriginal peoples did long ago in Mesoamerica in propagating those cacao trees that produced the most seeds. Perhaps these early people did not understand budding or grafting, but instead planted seeds on a select basis, choosing seeds from certain trees and not from others. Then, like now, there is little knowledge about a "pure line" or genetic lineage in cacao. Ultimately the successful cultivation of cacao, as in any crop, is highly dependent upon a knowledge of genetic characteristics and the ways in which they affect or influence relevant agronomic or agricultural features such as crop yield, hybrid vigor, resistance to disease, and growth. In spite of a long history of cultivation, relatively little is known about the genetics of the cacao tree—a situation aggravated in large measure by the long forty-year-plus life span of individual trees and, therefore, the multi-year time periods between successive generations of breeding. Such a situation precludes the gathering of genetic changes in successive generations of cacao. Thus, in spite of cacao's obvious importance in the evolution of the culture, economy, social well-being, and religion of ancient peoples in Mesoamerica and perhaps in South America, and its long history as an important commodity in modern times, there has not been the degree of understanding of cacao genetics and breeding as there has been of other crops such as maize, a food staple. But ever since the Maya Indians, who surely noticed the high degree of variability between pods and their seeds in different trees and must have begun to select varieties of *criollo* cacao long ago, human beings have tinkered with the genetic mystery of cacao—not knowing quite what they were working with.

In more recent times, open crossing among varieties in plantations has altered the degree of purity within many varieties, since pollinators transfer pollen between varieties, bringing together new genetic combinations. Thus some of the altering of varieties has been due to natural events. Of particular interest in this context has been Matina cacao, a type of *forastero* cacao found in Costa Rica, today in the Atlantic lowlands near Limón. Matina cacao has pale green, melon-shaped pods resembling those of *cacao común* in Brazil. Matina cacao is a testimonial to the birth of Costa Rica's cacao industry centuries ago. In the town of Matina alone, an estimated 200,000 cacao trees were in cultivation in 1736 (Patino 1963). And in 1920 the cultivation of cacao greatly expanded when the United Fruit Company replaced bananas with plantations of Matina cacao. Surprisingly, Matina cacao has remained fairly unadulterated into modern times in the Atlantic region of Costa Rica, even though many plantations were abandoned because of *Monila* disease in the early 1980s.

In some plantations, there is an estimated 10 percent erosion of pure Matina cacao owing to hybridization with a low proportion of *trinitario* cacao. Matina cacao, under the names "Cacao Costa Rica" and "Cacao Ceilan," was introduced long ago into Mexico and Guatemala, where it was crossed with local *criollos* and other types. And here again, as in Costa Rica, there has been little genetic erosion of Matina cacao: Pure or close-to-pure plantations of it exist today. This is a curious anomaly considering that many other types of cacao, when cultivated, typically hybridize to yield a cornucopia of still other types.

Cacao, Research, and the Global Market

A golden era of cacao research, focusing on disease control and resistance, took place in the 1950s and early 1960s. Research programs came to life not only in Costa Rica but in Brazil, Ecuador, Trinidad, and elsewhere. Field trials were set up in Guatemala, Costa Rica, Haiti, Colombia, Peru, Ecuador, and Bolivia (Hunter 1961). In these studies, a standard variety, ICS-95, was used as a benchmark against which targeted local varieties were then assessed for yield. Much of this work led eventually to the publication of a compilation of agronomic and other features of hundreds of cacao varieties, called the *Cacao Cultivars Register* (Enríquez and Soria 1967).

The field trials across Latin America were abandoned in the mid-to-late 1960s, when the world market price for cacao beans rose sharply, to an eventual high of more than $4,000 per metric ton by the late 1970s. Emphasis in agronomic studies and associated extension work shifted to increasing the planting of cacao, regardless of whether or not particular varieties had been field-tested. Increased planting, of course, typically leads to an oversupply later. Thus when prices began to drop after 1978, growers began abandoning cacao and planting other crops. At this time in Costa Rica alone, cacao production dropped from about 35 million pounds of beans in 1978 to about 4.5 million pounds in 1990 (Hunter 1990). Such drastic, long-lasting dips in agricultural production have more to do with local or regional social, economic, and political factors in cacao-producing countries than with disease and pestilence.

A case in point about the interplay of many factors affecting cacao production is Trinidad. From the beginning of the twentieth century until World War II, the tropical island produced as much as 10 percent of the world's supply of cacao beans—a hefty 75 million pounds of high-quality cacao beans a year. Much of the impetus for this phenomenal production came from Great Britain, which controlled Trinidad as a British colony until the middle of the 1900s. Through the Imperial College of Agriculture and, later, the University of the West Indies, the British sent their scientists to Trinidad to develop crop improvement research programs aimed at enhancing the cacao industry. But there was relatively little training of native Trinidadians along the way. Therefore, the role of research in improving cacao production did not take hold in the local population until recently.

The combined effect of a lack of training for Trinidadians and the resulting tardiness of their direct participation in cacao improvement research programs was a major factor in the dramatic decline of Trinidad's cacao industry. Trinidad's current level of production is between 2 million and 3 million pounds of cacao beans a year—yields of only 200 to 300 pounds per acre each year (Hunter 1990). At the same time, Trinidad produces a highly desirable, fine-flavored bean, in great demand on the world market.

According to some observers, there is little incentive to plant cacao on the island today. But the potential exists to bring back the cacao industry in Trinidad, given its long history of cacao production. The interest of the business community, the governmental holdings of experimental cacao plantings, including clonal gardens, and the active research program of the Cocoa Research Unit in the Faculty of Agriculture at the University of the West Indies hold great promise for the present and future.

An Ancient Crop in Modern Times

Several U.S. chocolate manufacturers today have their own cacao plantations in the American tropics, although worldwide, only 20 percent of the total production of cacao comes from large-scale modern plantations. Hershey Foods Corporation of Pennsylvania operated the Hummingbird cacao farm in Belize for many years. Although the operation was closed a few years ago, the goal was to introduce the cultivation of cacao to the country and provide a model farm for other growers. The M&M Mars Company of New Jersey operates the Almirante Cacao Research Center in Bahia, Brazil. The World's Finest Chocolate Company of Chicago operates the Union Vale Cocoa Estate, a 238-acre cacao plantation on Saint Lucia. Union Vale, although owned by the chocolate company, is also a member of a local agricultural cooperative, to which it sells its cacao crop. World's Finest then buys back the beans at the going market price.

Cacao farming can be a costly endeavor, with start-up costs of $3,000 or $4,000 an acre, the use of chemical fertilizers and pesticides to obtain good yields, and the labor of pruning the trees and hand-pollinating the flowers. And tree crops like cacao represent a long-term commitment for the farmer. Still, the vast bulk of the world's cacao is produced by small farmers. These farmers depend in large measure upon government-funded cacao breeding and extension programs to ensure that they have adequate supplies of planting material for their plantations.

National or multinational government agencies in cacao-growing countries, often with some financial support from private groups such as the U.S. Chocolate Manufacturers Association, may provide programs of cacao breeding and disease control to assist small farmers or growers. Efforts have also been under way to help small farmers of cacao diversify into other tree crops, such as coconuts, which may require less care.

Wherever cacao is grown, farmers are faced with the challenge of obtaining an economically suitable yield of beans from the plantation. Under reasonably good soil conditions, a single mature cacao tree may produce hundreds of pods in a year, although thirty or forty is more the average. This challenge has not changed appreciably over the long history of cacao cultivation.

What also has not changed much is the manner in which cacao beans are harvested. Harvesting the pods is hard physical labor, one without mechanized shortcuts. Twice a year, mature pods are cut from the trees with a sharp machete. The pods are collected in sacks and dumped into a pile, where eventually they are cracked open and the seeds are removed.

The process is slow and laborious. A worker sits on a crate in the shade and cracks open the tough pods with a sharp cutting tool or machete. A deep, crosswise cut is made into the pod, which is then broken in half. The glistening mass of seeds gets scooped out of the pod and dumped into a deep wooden box. The empty pods are tossed into a pile on the ground and left to rot. The fermentation of the gathered seeds is usually done in the same sturdy wooden boxes used in harvesting, although the receptacle used often varies in different parts of the world. This is usually accomplished in a covered wooden box, sometimes using burlap or banana leaves as the cover.

The cacao beans usually undergo several days of fermentation in which anaerobic microbes feed on the energy-rich mucilaginous pulp coating the seeds, converting the pulp to alcohol, as in the production of wine. During fermentation, polyphenols in the seeds are converted into the chemical precursors of the coveted chocolate flavor, and the outer seed color changes from reddish tan to purplish. Fermentation kills the cacao embryos while microbial action on the outside liquifies the pulp encasing each seed. As the process proceeds, the astringency and bitterness of the seeds are lessened, and the chocolate flavor is enhanced. *Criollo*-derived cacao seeds require less time to ferment than *forastero* seeds, possibly due to the higher quality of chocolate flavor found in *criollo* cacao.

As sugars in the pulp and seed coat are used up, air enters the seed, and a second fermentation, lasting several days, is set into play. This aerobic process converts the alcohol into acetic acid and vinegar. This is the same process that eventually converts wine to vinegar when it is left open. The high temperatures generated in the fermentations and the penetration of the seed with alcohols

A worker at the La Lola Experimental Farm cutting open ripe cacao fruits and dumping the white, pulp-coated fresh seeds into a wooden box that fits over the back of a mule or small horse. From various points of harvest in the plantation, the collected seeds are moved in this manner or by tractor-drawn wagons to a central area, where they are fermented and dried.

A small horse outfitted with a wooden box and two gunny sacks for carrying harvested fresh cacao seeds from the plantation to the seed fermenting and drying area. Shown with the horse is Miguel Cerdas, a long-time manager of the La Lola Experimental Station.

and acetic acid not only kill the embryo but also induce a series of incompletely understood chemical changes that produce the chocolate taste and aroma. The beans, now changing from purple to dark brown, become less astringent. At the same time, water loss shrinks the seeds from inside the seed coats.

Following fermentation, which takes place either at the plantation or a nearby fermentary, the beans are dried in the sun or with a fire or fuel-generated dryer, further reducing their moisture content to about 7 percent, which prevents mold formation during storage and shipment to chocolate factories. Before they are shipped, batches of dried and sorted seeds are checked for mold or fungal infestations and tested for flavor, aroma, and size. At this stage the beans give off a pleasurable smell and have a slight chocolate flavor, but they are still bitter and oily.

Part of the mystique of cacao is that its production still largely depends on the small-scale farmer, deploying fairly ancient methods for curing the beans, while the technology for processing chocolate has been large-scale in response to high

Drying cacao seeds in the sun at La Lola. The seeds are periodically turned with a rake to ensure evenness of drying.

A cacao plantation worker's house near La Lola. To the left of the house is a small cacao seed drying table with a corrugated sheet metal roof. This family may own or rent a small plot of cacao and process the seeds themselves before selling them.

demand. The beans are bought in mammoth quantities by chocolate companies and candy and confectionery manufacturers, usually through international brokerages or exchanges situated not in the tropics but in the industrial centers of the Western world. The dried beans get shipped to chocolate companies all over the world.

The modern manufacture of chocolate involves a series of processes resulting in the production of chocolate, cocoa powder, and cocoa butter (Wood and Lass 1985). At the processing plant or factory, the seeds are cleaned of any remaining foreign matter and then roasted in 121-degree-C (250-degree-F) ovens for beans that will be used in making chocolate, and slightly higher temperatures for beans to be turned into cocoa powder. The roasting removes water and acids, leaving the desired chocolate flavor. Roasted seeds are then cracked to free the large cotyledons, and the shells or husks are removed. The shells are commonly used for a fragrant garden mulch or get pressed to yield cocoa butter. Or the husks may be used to extract theobromine, which can be chemically converted into caffeine for use in various beverages and medicines.

After the beans have been roasted and cracked, the separated cotyledons are called nibs, the material used to make chocolate. Grinding the roasted nibs produces chocolate "liquor," which has a fat content of 55 to 58 percent. The liquor can be molded into small squares to make baking chocolate or turned into other products. Until the middle of the nineteenth century, chocolate beverages were made by dissolving these blocks of chocolate in hot water or milk, the latter an English innovation, to which sugar was added (Chapter 2).

To make chocolate today, chocolate liquor is processed in hydraulic presses to remove some of the fat (cocoa butter), resulting in a cocoa powder with 22 to 23 percent fat content. Further pressing yields a cocoa powder with an even lower fat content. The high-fat cocoa powders are used to make drinks; the low-fat powders, containing 10 to 13 percent fat, are used as flavoring in cakes, ice cream, and other food products. Plain chocolate is made by mixing the liquor with sugar and enough cocoa butter to mold the chocolate into bars. In the United States milk powder is commonly added to the liquor to make milk chocolate. In Britain and some other countries milk chocolate is made by first condensing fresh milk with sugar, then adding the chocolate liquor.

Chemists still do not understand the nature of the chocolate flavor, which depends upon hundreds of substances in the cacao seed or cocoa bean. Attempts to manufacture an acceptable synthetic chocolate, bypassing the cacao tree altogether, have failed. So although chocolate is to some extent an invented product in that the seeds are treated in certain ways to produce the unique flavor, no one has successfully come up with a synthetic counterpart to the chemistry wrought in the rain forest. The curing of the cacao seed, the first step in the production of chocolate, depends upon the biological and chemical material provided by *Theobroma cacao*, the living tree and its complex natural history.

Excursions into the Natural History of Cacao and Cacao Plantations

Much of what I have come to know and appreciate about the natural history of cacao is based on my personal experiences and observations in the Caribbean watershed region of Costa Rica, where, as in other parts of the tropics, patches of cacao, bananas, and other crops sometimes intermingle with tropical rain forest. In some places the plantations—small *fincas*—are enclosed by rain forest. In other places the rain forest is much less evident, and cacao plantations may alternate with expansive banana fields, coconut groves, and swatches of secondary forest. But throughout the sprawling Caribbean flank of Costa Rica, cacao has been an ancient crop, embedded not only in the landscape of rolling foothills and river-streaked plains, but also in the culture and history of the region. The natural history of the cacao tree is woven within the ecological fabric of the rain forest.

Because they are usually planted in low-lying areas, trees in cacao groves often appear smaller than their actual height of 3 or 4 meters. The trees can be recognized almost immediately, however, by their leafy crowns frequently streaked with new flushes of bright red foliage. The brilliant, gaudy flushes of new leaves are due to the presence of greater amounts of red pigments, the anthocyanins, than in the mature leaves, which are always dark green. Cacao's leafy canopy, dark branches, and lichen-spotted trunk render it fairly inconspicuous against the backdrop of rain forest. But it is the strange-looking, glistening fruit that gives away its presence and pulls you in for a closer look.

At Finca El Uno, in Sarapiquí, Costa Rica, tall, stately rubber trees, 50 feet or more in height, form a dense, shaded woodland, giving the farm an almost forest-like appearance. The rubber trees are planted close enough together to create a canopy, providing considerable shade for the neatly planted rows of cacao underneath. The manner in which cacao is grown here is representative of how it is often cultivated and farmed throughout much of the wet tropics.

The sometimes dense shade in a cacao plantation makes for a thick, moist carpet of leaf litter and other plant debris—a layer of nutrient-laden mulch important to the health of the cacao and its overstory. This matted tangle of fallen leaves, cacao pod husks, fallen branches, and more is what keeps the plantation alive and healthy. The squishing mat of nature's decay is also evident in the rain forest. Above the mulch layer, within the tangle of blackish branches of the cacao trees, thin wisps of moss, lichens, and tiny bromeliads often cling to the branches; much higher up, giant bromeliads hug the trunks and branches of the rubber and other trees. At certain times of the year, *Maxillaria* and other orchids bloom from their aerial perches on the rubber trees.

The interior of a cacao plantation can be inspiring. Consider this poignant description of a cacao farm in Trinidad by the novelist and native of Trinidad, V. S. Naipaul (1981): "The cocoa woods were another thing. They were like the woods of fairy tales, dark and shadowed and cool. The cocoa-pods, hanging by thick short stems, were like wax fruit in brilliant green and yellow and red and crimson and purple. Once, on a late afternoon drive, to Tamana, I found the fields flooded. Out of the flat yellow water, which gurgled in the darkness, the black trunks of the stunted trees rose."

I, too, have experienced the mystique of old cacao plantings. There is something wonderful about a plantation of wizened old cacao trees. Within this setting, moisture drips from every leaf and branch, and the mulch smells steamy and fresh. Little pools of sunlight filter through the large, flat leaves of the cacao trees, illuminating the leaf litter with its earth tones of russet, yellow, orange, and brown. The dapple of sun and shade often traps for brief moments the to-and-fro passage of flying insects, many unknown and certainly unnamed by science, just above the leaf litter floor. A rich variety of ants, grasshoppers, and other insects

Full-sized, ripe fruits of *Theobroma cacao* L. Immediately below the central fruit in the picture are the drooping flush leaves of a typical chupon, or sucker (shoot), growing out from the base of the tree trunk. The two smaller fruits are cherelles.

thrive on the cacao tree and in the leaf mulch (e.g., Leston 1970; Young 1983a, 1983b, 1983c, 1983d, 1984a, 1984b, 1986a, 1986b, 1986c, 1986d; Young et al. 1987b), as do small vertebrate predators, especially *Anolis* lizards (Andrews 1979).

Flowers

Against the blackish branches of the cacao trees, the flowers stand out as small bold white stars. The flowers often break through the thick mats of mosses growing on the trees, along with tiny ferns and seedlings. One is struck by the large numbers of flowers on the trunk and along the branches of the tree, a habit known as cauliflory, rather than at the terminal ends of the branches.

Cauliflory, as proposed by P. W. Richards and others (Richards 1952), may be

Flowers of *Theobroma cacao* L. Just below and in front of the expanded sepals, note the inwardly curved petal pouches concealing the anthers, or pollen sacs. Note also the long dark staminodes, forming a picket fence around the whitish pistil. The tip of each petal bears a recurved, flattened ligule, which may provide pollinating insects with a landing platform. An unopened, mature floral bud is seen to the left of the three open flowers in this particular inflorescence.

an adaptation to ensure pollination in the shaded understory of the rain forest. Ants and small flies abound here, and these insects, rather than bees, may figure strongly in the pollination of well-shaded cauliflorous plants. Although cauliflorous species are also pollinated by bees, wasps, bats, and other animals, the arrangement of flowers on the trunk and branches in the heavily shaded rain forest understory might facilitate pollination by small, dryness-intolerant insects, especially ants and flies associated with leaf litter.

Also striking is the proclivity of cacao trees in most plantations to produce many flowers, each one a dime-sized pentagonal configuration of predominantly white petals, with reddish streaks on the petals that are only apparent up close. Equally impressive, too, is the preponderance of fallen blossoms, forming a sprinkling of white patches on the otherwise earthen-hued leaf mulch beneath the trees. The white patchwork underfoot attests to one of the biggest biological mysteries about the cacao tree: Why do only 1 to 5 percent of each year's flowers produce fruit?

Pollination biologists offer three alternative explanations of marked disproportionality between flowering and fruit production (e.g., Stephenson 1981): pollination is the limiting factor, and pollinators are scarce; resources are insufficient to support more than a low number of developing fruits; or both flowers and fruits succumb to disease or other factors. In *T. cacao*, pollination turns out to be the limiting factor (Chapter 5).

Certain other species of *Theobroma* and *Herrania*, the "wild cacaos," provide clues to interpreting the ecological significance of cacao's floral design and habits. Although largely unstudied in terms of their natural history, these species show a tremendous range of adaptations, often specialized, for animal-mediated pollination. Some Sterculiaceae in the neotropics, such as *Theobroma grandiflorum* in Brazil, have large, showy flowers attractive to certain kinds of bees, such as the stingless meliponines (Aguiar-Falcao and Lieras 1983). Others have cream, pale-green, or purplish-colored clustered flowers, with musty fragrances, suggestive of bat pollination. Still others, such as *Sterculia chicha* in Brazil, have pale, greenish-white and reddish flowers pollinated by stout-bodied flies belonging to several families, such as Sarcophagidae, Muscidae, and Fannidae (Taroda and Gibbs 1982).

What is also interesting about species of *Sterculia*, the nominal genus of the family (named after Stercus, the Roman god of carrion and dung) is that the flowers, typically in inflorescences near the tips of the branches (not cauliflorous as in many species of *Theobroma* and all species of *Herrania*), exude a pungent odor of rotten meat that attracts flies as pollinators (Pijl 1953; Young 1984e). Unlike *Theobroma*, *Sterculia* species are giant canopy trees of the tropical rain forest, and their seeds, visible when the large, red pendant fruit split open, are dispersed by monkeys and large birds such as parrots (Taylor 1987). Botanical descriptions of *Theobroma cacao* are known from as early as the 1600s, but the first full written description of this species is recognized to be that of Linnaeus in his *Species Plantarum* (1753), which was based upon specimens of a tree collected in Jamaica. Linnaeus described four prominent genera within the Sterculiaceae: *Theobroma*, *Herrania*, *Guazuma*, and *Cola*.

The result of such evolutionary divergence in flower-pollinator interactions among closely related species within a genus, or among allied genera, especially within the same geographical region (e.g., lowland Amazonia), is an eventual increased species richness of animals associated with plant groups. This is a basic lesson of the tropical rain forest, one also applicable to herbivorous species, symbionts, and seed-dispersers associated and evolving in conjunction with certain groups of plants.

Floral Design

Theobroma and *Herrania* display a similar pentamerous design in floral structure. Five petals, usually curved into pouchlike structures, conceal five anthers peripheral to the centrally located stigma. Between the petals and the stigma occur five staminodes, usually taller than the stigma, and arranged alternately with the petals. In *T. cacao* and several other species of *Theobroma*, the staminodes form a circular fence enclosing the stigma and seemingly act as a mechanical barrier between the stigma and the distantly situated anthers, which are partially concealed in the petal "pouches." This floral design is suggestive of a breeding system promoting outcrossing—cross- fertilization between different trees within a species—since it appears mechanically difficult to transfer pollen from the anthers to the stigma of the same flower. In addition, many forms of cacao are genetically self-incompatible, meaning that pollen from one flower cannot successfully fertilize the same flower or other flowers on the same tree.

The petal in cacao starts out narrow at the base, expands into the pouch, and ends distally with a broadened and flat lip of tissue, the ligule. Whereas most of the petal is white with reddish markings, the ligule is often yellow, a conspicuous structure to the human eye. The upwardly flexed posture of the five petal ligules on a freshly opened cacao flower render the entire flower more visible against the usually muted shade of the foliage.

Ten stamens, arranged in two whorls, form the androecium (male portion) of the flower. The outer whorl consists of five nonfertile red staminodes, the inner

Inflorescences of *Theobroma speciosum* Willd. Note also the ripe, spherical fruit.

of five fertile stamens. Each stamen bears two anthers or pollen sacs concealed inside the pouch of the corresponding petal. The single, centrally located style is about twice the length of the ovary, which contains many ovules arranged around a central axis.

The yellowish pollen is very sticky and often remains lodged inside the petal pouches following dehiscence (splitting of the pollen sacs) of the anthers. Depending upon the variety, cacao flowers range from pinkish to white, and most have red-pigmented lines running lengthwise on the inner surface of the petals. Some students of cacao pollination believe these are "nectar guides," even though the flowers do not appear to produce nectar (Young et al. 1984). The flowers are weakly fragrant to the human nose, even though floral fragrance is certainly present. Old leaf axil scars become inflorescence "cushions," with about a month's time required for floral buds to push through the bark and mature. Floral cushions become thickened swellings after a tree has been producing flowers for several years.

Flowers begin to open very slowly late in the afternoon, continuing through the night until they are fully open by about 5:00 A.M. the following morning. Shortly thereafter the anthers release pollen, which remains viable for one or two days. The receptivity of the style and stigma is greatest on the first day of flowering, the best time for pollination. Flowers which are not pollinated—a surprisingly common event—fall off the tree by the end of the second day.

A conspicuous feature of this basic floral design is the presence of the petal ligules. The various species of *Theobroma* and, to a lesser degree, *Herrania*, exhibit dramatic diversification in the shape, size, and color of their ligules. In some *Herrania* species, the ligules are elongated into ribbon-like appendages that flow gently in the breeze and aid in pollination (Young 1984e). Stout-bodied phorid flies alight on the swaying ligules, crawl up them, and enter the flower, where they pick up large quantities of pollen on their bristle-covered bodies. In rummaging through the flower, a phorid brushes pollen onto the pistil and stigma, resulting in pollination (Young 1984e).

Fruit

By most botanical standards, the pod of the cacao tree is a large fruit. Each pod contains thirty to forty large seeds, each about the size of an almond. Young fruits are greenish or maroon and may be difficult to spot in the branches and leaves, their colors blending with the muted shadows and sun flecks of the habitat.

The new pod grows slowly for about forty days following pollination, at which point the zygote, formed by the fusion of an egg cell with a pollen grain, undergoes its first cellular division. The embryo, or future cacao seed, then grows rapidly for the next seventy-five days. Between five and six months are required for the complete maturation of the pod and its complement of thirty to forty seeds, neatly arranged and centrally attached to a fibrous placenta. Mature pods often turn bright yellow or red and usually take an elliptical form, with either a smooth or ridged surface.

Even though typically only a small fraction of flowers on a cacao tree are successfully pollinated, enough new fruits are often set to invoke a thinning process called "cherelle wilt." Cherelles are immature pods that shrivel up, die, and turn black but do not drop off the tree. Wilted cherelles are sometimes mistaken for young pods killed by fungal disease, especially "black pod," caused by *Phytophthora*.

To sprout, the seeds must be unencumbered by the pulp. The mucilaginous pulp coating the seeds is believed to contain a "germination inhibitor." Once the pod is opened and the pulp begins to decay, thereby knocking out the germination inhibitor, the seeds, even those still inside the pod, begin to germinate.

Whether it is inside an opened pod or on the ground, a germinating cacao

Herrania flowers. Note the extremely long, ribbonlike petal ligules, which, together with a pronounced smell of rotting meat produced by the flowers, lure pollinating phorid flies.

seed first sets forth a tiny rootlet, while the growth of the hypocotyl, or stem shoot, arches up, raising the still-closed cotyledons an inch or so above the ground. The cotyledons then open, exposing four tiny leaves at the same level. Subsequent growth results in a spiral arrangement of new leaves on the seedling. The third phase of growth entails a remarkable arrangement of sideways-stem growth, the jorquette. The jorquette is the terminal end of the stem, where five buds grow out sideways at the same time, forming side or fan branches arranged at a 0–60-degree angle to the horizontal. Growth of these "fan branches" continues horizontally as the main stem continues to grow vertically. If there is damage to the apical growing point of the seedling's main stem prior to jorquetting (sending forth the first whorl of primary branches), buds below the top grow in an upright or semivertical to fully vertical position and have a spiral leaf arrangement. These secondary shoots, called chupons or suckers, also grow from the base of the tree's trunk and can replace the main trunk if the tree falls over. Budwood or cuttings taken from a chupon produce trees that grow upright, while those taken from fan branches result in trees growing sideways.

In their natural growth cycle in the rain forest, undisturbed chupons grow to exceed the height of the tree's original canopy, and their canopies replace lower ones. This process of chupon canopy replacement can result in a wild tree reaching 20 meters or so in height. Cacao trees in plantations reach 4 to 10 meters in the course of their twenty-to-forty-year life span, depending in some measure upon the degree of pruning, among other factors.

Flushing and Flowering

Periodically, and on a more or less regular annual cycle, the cacao tree puts forth "flushes" of new leaves: the terminal leaf buds on a branch quickly grow out, producing three to six pairs of either pale green (more typical of wild trees) or red soft leaves, which eventually harden and turn dark green. Following a flush cycle, new terminal leaf buds remain dormant for a while, until conditions bring about a later flush. Thus cacao growth is intermittent, with bursts of leaf flushing alternating with periods of vegetative rest. Depending in large measure upon local conditions, it is not unusual for cacao trees to produce two or three rather intensive flushes, or new shoot growth, during a year. Smaller bouts of flushing with an interval of three or four months also occur.

One of the most beautiful features of the cacao *fincas* in Costa Rica, and elsewhere in Central America, is the mixed array of colors of the cacao canopy—generous blotches of new-leaf crimson embossed upon the otherwise dark green

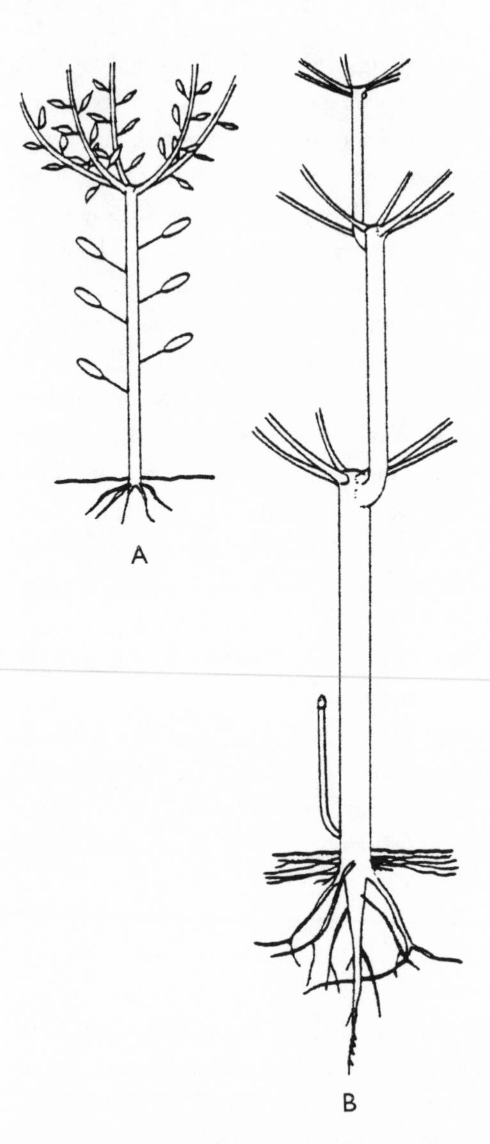

Stem growth of the cacao tree: (A) adult seedling with jorquette and five fan branches, (B) older tree of three stories (branching levels) and basal chupon. Modified slightly from Wood and Lass (1985).

sea of leaves stretching along the tropical rain forest. Why young leaves of cacao and many other rain forest tree species are red is unknown, although several hypotheses have been put forward (Levin 1971). One current view places little adaptive significance on the phenomenon, suggesting that red-producing anthocyanins are a by-product of the production of other substances in rapidly growing young leaves (Lee et al. 1987).

The control of the flushing process is of considerable interest. Often referred to as a "change of leaf," it involves a draining of nutrients from older leaves into new leaves. According to P. de T. Alvim, the flushing cycle in cacao is triggered by decreased rainfall and enhanced moisture stress, causing leaf abscission to occur (Alvim 1977). Leaf-drop breaks the dormancy of new leaf buds on the tree's branches because of diminished concentrations of the dormancy inhibitor normally found in the leaves of cacao. Flushing ensues quickly after more-or-less synchronized leaf-drop (although not all mature leaves drop off the tree, for cacao is an evergreen), immediately following an intense dry spell at the onset of rains.

Because cacao trees in plantations with maximal direct exposure to the sun experience moisture stress more quickly during dry periods, flushing in these open-grown trees tends to be more intense than in partially or heavily shaded trees. Generally leaf flushing in cacao is least during the rainiest times of the year. Flushing, as I have seen it in Costa Rican cacao *fincas*, is the cacao tree's changing response to the annual cycle of wetness and dryness that prevails in the Atlantic zone. Because of the mobilization of nutrients during flushing, this process also participates in the regulation of the tree's flowering and fruiting cycles, which also require shunts of energy away from other functions.

Flowering periodicity is also a highly structured phenomenon in cacao. Cacao trees in plantations generally begin to flower after three to five years, depending upon the local conditions and variety. Flowering follows a definitive seasonal cycle, with a big peak or burst of activity during the first rains after the dry season, even in only weakly seasonal localities such as the lowland tropical rain forests in southern Central America along the Caribbean coast. Yet some level of flowering typically occurs on most cultivars throughout the year. Flowering patterns of truly wild cacao trees in Amazonia have not been studied, but flowering might be less cyclic there owing to the position of the trees in the heavily shaded rain forest understory. Flowering in plantation cacao in Costa Rica tends to be lowest during the rainiest months of the year, and low points in flowering correspond to the times of heaviest fruit (pod) loads under these cultivated conditions. *Fincas* commonly carry out two harvests of pods annually, and a maturing crop of fruit places an energy drain on trees, which in turn dampens flowering intensity. Fruits compete with flowers for tree nutrients, especially carbohydrates. Thus flowers tend to be most abundant on trees in orchards of *fincas* at times of the year when pods are scarce or absent.

It therefore appears as if both external and internal cues regulate the flowering cycle in cacao. The tree's ability to set its flowering level, followed by eventual pollination and the maturation of set fruit, is a fine balance of limited energy reserves shunted to different functions. This energy store is shaped by the tree's photosynthetic sunscreen and metabolic requirements, the canopy of leaves, and trace minerals and nutrients garnished from the mulch and soil below it.

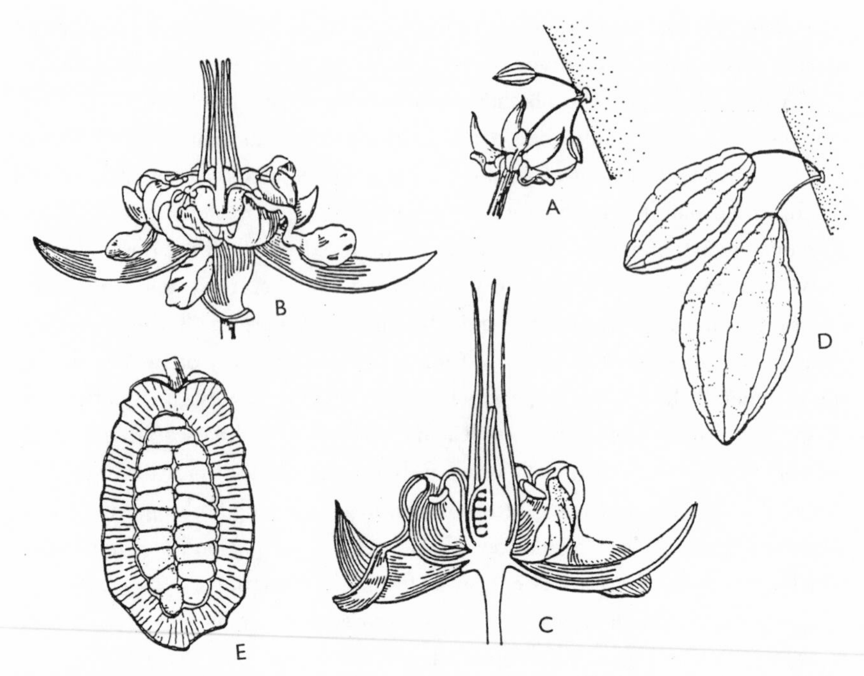

Flowers and fruits of the cacao tree: (A) cauliflorous flower, (B) flower greatly magnified, (C) cross section of a flower, (D) fruits borne directly on the trunk, (E) cross section of a fruit. Drawing by M. C. Ogorzaly, reproduced with slight modifications from *Economic Botany*, by B. B. Simpson and M. C. Ogorzaly (McGraw-Hill, 1986). Reproduced with permission of the authors and McGraw-Hill.

Soils and Mycorrhizae

The ecology of soil and mulch in the tropics, especially of fungal and microbial associations, is a pioneering field of research, one that has yet to elucidate all the complex interactions of the microorganisms of the soil with invertebrates and plant structures along the forest floor in both the temperate zones and the tropics. The ecological impact of cultivating rain forest tree species on these complex associations between tree roots and microbiotas warrants considerable agronomic attention as well.

Old fallen branches, pod husks, and other cacao tree debris create a damp, shaded niche on the forest floor that is a hotbed of fungal activity. This dense, soggy mulch, rife with decaying organic matter, is the ecological setting for the life-sustaining theatrics of the mycorrhizal fungi that live within the tree roots.

The root system of cacao is well-designed to accommodate the mycorrhizae. A sturdy taproot up a meter or more in length has many lateral feeder roots networking through the upper inches of the topsoil. It is the tips of these lateral roots that are coated with brushes of fine rootlets, especially where there is a lot of decaying matter such as mulch. These are the sites for mycorrhizal action.

When the tentaclelike hyphae of these fungi invade the surface or near-surface, they promote the proliferation of rootlets that spread out, in tentacle fashion, across the forest floor in search of minerals and other nutrients. The presence in a mature cacao tree of many feeder roots coated with rootlets and fine hairs enables the tree to participate effectively in the mobilization of nutrients.

Mycorrhizal fungi living in symbiosis with its roots permit the cacao to absorb minerals such as phosphorus from the upper layers of the soil and mulch that are otherwise unavailable to the tree, in exchange for waste products from the tree, which are needed by the fungi. The uptake of phosphorous, for example, intermediated by mycorrhizae, enables plants to synthesize nutrients. In exchange for this invaluable service, the fungi take up carbon from the tree. This is important because fungi cannot photosynthesize their food. It is this mutually beneficial partnership between tree and fungus that sustains many kinds of trees in the tropical rain forest.

Newly bedded tree seedlings, including cacao, have a patchy opportunity to become infected with beneficial symbionts such as mycorrhizae (e.g., Alexander 1986). Fungal spores, the reproductive propagules, are dispersed by rain and mulch-inhabiting arthropods or other animals, although there is a dearth of information on the nature of their role. The activity of such organisms likely determines to some degree the spatial and temporal structuring of mycorrhizal fungal-host tree relationships. When the landscape is stripped of rain forest, the mulch and topsoil sometimes lose their microbial and fungal tree symbionts.

Although it was discovered long ago that the feeder roots of cacao possess mycorrhizal fungi (Laycock 1945), it was not until the studies of David P. Janos at Finca La Selva in Costa Rica that the important role of these fungi in mineral uptake from the soil was demonstrated experimentally (Janos 1980). In an abandoned cacao plantation along the Río Puerto Viejo on the La Selva property, Janos discovered that fallen branches from cacao trees, left in the deep mulch, developed large cushions of mycorrhizae, which he then used as a source of inoculum to examine the effect of mycorrhizae on the growth of the rain forest seedlings. The inoculum resulted in mycorrhizal infections taking hold in sixteen out of twenty-four plant families tested.

In older cacao plantations, with partial shade cover, the ground cover is likely strewn with patches of these fungi associated with the superficial feeder rootlets, and other cacao debris. In the rain forest proper, the large spores of mycorrhizal fungi are relatively rare, and the colonization of new trees is perhaps a chancy matter. Mycorrhizae place a hefty demand on their tree hosts for carbon-based minerals, which are largely unavailable or otherwise inaccessible to absorption into tree rootlets (Janos 1980).

Trees possessing mycorrhizal fungi may have a competitive advantage over those lacking them under conditions of restricted soil nutrient supplies. The root systems of lowland rain forest trees therefore participate in intricate symbiotic interactions of these plants with certain kinds of nutrient-mobilizing fungi and bacteria—perhaps even more elaborate and common than what is typically encountered in temperate zone forests. Since Mayan times, leguminous tree species have been planted as shade cover over cacao trees, greatly augmenting the flux of mulch and soil nutrients available to cacao.

There is some evidence in other plant species that mycorrhizal infection of cultivars greatly increases flowering duration, reduces the time for onset of initial flowering, and increases seed production and the proportion of set flowers producing mature fruits (Alexander 1986). Mycorrhizal association of cacao cultivars in plantations may push up the trees' productivity for fruit production, assuming adequate pollination.

It might well be that mycorrhizae are generalist-opportunists in terms of the different species of trees they infect in tropical rain forest. Planting various species of trees within the cacao plantation, or leaving part of the overstory of canopy trees intact when planting cacao, might ensure the successful spread of mycorrhizae from tree to tree within the surface-level, mulch-covered networks of feeder roots. The chances are good that such inoculum would spread among the cacao trees when various species of endemic rain forest trees are present, since as with cacao, many or all of these could be obligate symbiotic hosts of the fungi.

In the rain forest, clumps of wild cacao trees, owing to their spatially scattered distribution, interspersed with many other kinds of trees, can garnish their share of the mycorrhizal layer. But in some plantation settings, out in the open, this added patina of natural fertility to the mulch and soil may not always be sufficient to feed all of the trees. When an overstory of rain forest canopy trees, or subcanopy trees, is left to provide shade to the cacao trees in the plantation (Young 1982), a reservoir of soil and mulch biotas can be established that may naturally fertilize the cacao grove. An inoculum of mycorrhizae may spread from one or a few trees in the plantation among many, thereby helping to reestablish conditions suitable for the growth of cacao trees. Keeping a certain amount of shading foliage over the mulch layer in the plantation, and thereby ensuring its moistness, may help to create these conditions.

Janos (1980) maintains that big-seeded tree species, like cacao, which are typical of primary-growth or mature tropical rain forest, have an obligate dependency upon mycorrhizae. Being large, they have a higher demand for growth once the cotyledons expand and thereby become attractive resource patches for colonizing mycorrhizal fungi. Large stores of food allow the seedling to initiate growth and to attract and accumulate a sizable colony of fungi, which then takes hold and becomes established on the fledgling root system. Large-seeded tree species, as seedlings, have a greater demand for mineral uptake than smaller-seeded species in an environment generally deficient in minerals. Therefore, one expects tree species such as *T. cacao* to have an obligate, rather than facultative, symbiosis with mycorrhizal fungi.

According to Janos, the heavily shaded cacao trees in their natural state in the rain forest understory might shunt limited supplies of photosynthate to feed their mycorrhizal fungal colonies, thereby restricting the amount of this food available for the trees to produce flowers and fruit, once they are mature. But in the relatively less shady conditions of the plantation, as in a typical cacao *finca*, cacao trees undoubtedly fix greater quantities of photosynthate, making available a greater supply of carbon-based food for both the tree and its obligate symbiotic fungi. Under these less constraining conditions, the cacao trees flower more profusely and develop bushy canopy foliage. Furthermore, in the open plantation, cacao trees enhance their ability to maintain substantial colonies of the mycorrhizal fungi since more food, through photosynthesis, is available. The increased uptake in minerals out in the open, compared with the shaded forest understory, results in greater flowering and fruit maturation in cultivated cacao trees.

The implications of these assertions for cacao bean yields on plantation trees are very interesting. Out in the open, all else being equal, the cacao tree is predicted to be less conservative in its reproductive output, both vegetatively and sexually.

Animal Interactions

The beneficial role of mycorrhizal fungi on cacao trees in plantations might be offset or diminished by the increased stress placed on these trees by herbivorous insects. Cacao trees in plantations are routinely deluged with many kinds of herbivorous insects attacking the leaves, flowers, and pods. In Central America leaf-cutter ants defoliate cacao trees. The caterpillars of various species of noctuids, notodontids, and other moths attack cacao meristem and mature leaves (Lara 1957; Young 1983d, 1984a). Various leaf beetles devour the flowers (Young 1988b).

Stress from insect attack is most obvious on plantation cacao trees having minimal or no shade, and, overall, cacao trees on plantations are far more heavily infested than are wild trees in the tropical rain forest. This intense herbivore pressure may diminish the mutualistic interaction of the mycorrhizal fungi by lowering the amount of photosynthesis (through the removal of leaf surface area by insect feeding) and, therefore, reducing the energy that would otherwise be shunted to the fungi. Such an effect has been demonstrated recently in the pinyon pine of North America (Gehring and Whitham 1991). Reducing the energy available to the fungus may not only retard nutrient uptake from the soil, but also increase the vulnerability of the tree to pathogens. Under these conditions, the benefit of the symbiosis to the cacao tree would be weakened somewhat, even though there is still net benefit.

The intervention, by necessity, of ecological factors such as pollinating insects, disease-causing fungi, and other pathogens offsets the size of the crop yields away from the improved physiological capacity of the trees in the plantation setting. Under such circumstances, natural pollination, a function of the insects' access to the trees, especially at times of peak flowering, may be a limiting factor in determining harvest sizes. Pod diseases may be another such limiting factor.

The cacao tree's interactions with pollinators, herbivores, seed-dispersal agents, and symbiotic fungi in its wild state illustrate its close evolutionary ties to its natural habitat, the tropical rain forest. Many species of both plants and animals associate and interact with the cacao tree in its natural wild state. In this context it is no different than the myriad other plant species of the tropical rain forests.

During the lengthy rainy season typical of ideal cacao-growing localities in the humid or wet tropics, although it is a period of low flowering for the cacao, it is often the time of deepest leaf mulch accumulation, conditions also ideal for the proliferation of cacao-pollinating midges (Chapter 5), as studied in Brazil (Winder 1972, 1977; Winder and Silva 1975) and other localities. Ironically, times of greatest flowering, such as at the onset of the rainy season, are typically low points in the natural cycle of pollinator abundance in cacao cultivated in plantations.

During the wettest months, cacao plantations teem with scores of blackened, rotted pods hanging from the trees. These pods, soft and pliable, are ideal breeding substrates and hideaways for many kinds of insects and other arthropods, some of which figure strongly in the cacao tree's life cycle and natural history. Some of these creatures live within the spongy pod wall, while others occupy the rotted-out central cavity and seeds. The rotting pods are a wellspring of fungal, bacterial, and even vascular plant life, all providing a substrate for minute forms of animal life. Ants are by far the most commonly encountered occupants of arboreal rotting cacao pods in Costa Rica (Young 1986c).

Each kind of pod damage tells a different story of ecological dependency. Scrape away the tough, stringy pod wall with a pocketknife in an area where there is substantial discoloring from a fungus and one finds fly maggots. How the fungus attacks this particular pod in the first place is unknown. Perhaps, when it is still developing, an insect such as a shield bug pierces its skin to feed, at the same time opening a wound for easy entry by fungal spores floating in the air. Or perhaps spores splash onto the tiny hole, invisible to our naked eyes, in a rain drop washing off a leaf above the fruit.

Tap other blackened pods and stir up clouds of resting mosquitoes and other flies. These creatures, active at dusk and night, spend the day sitting on the fungus-lined shell of the pod. Sciarid midges, cecidomyiid midges, and other flies are also part of this consortium of creatures that breed in the dead pods. Most likely some of these midges are playing out their mating games in the nighttime airspace near this pod, ripe not for picking but for midge maggots. Mating will surely be followed, in accordance with the cadences of an instinctive script, by eggs being placed into the putrefying flesh of this cacao pod, ensuring once again a cycle of renewed life.

The cacao tree is a food source for many other kinds of insects, including the flower-eating leaf beetles *Colaspis* and *Monolepta* in Central America (Young 1988b). Petioles of cacao flowers are a favored feeding site for other insects, including aphids, the latter tended by ants. *Crematogaster* and other ants stroke the aphids to obtain a sugary, honeylike fluid, which the ants eat. Many species of ants nest in the dense cacao leaf litter or the soil beneath the litter. Some species of tiny ants nest deep in the soil and are seldom found by biologists.

Aphids are more prevalent on varieties of cacao with a vivid red pigmentation in the floral parts, including the petiole. While the reasons for this behavior are unclear, perhaps red-tinted flowers are more visually attractive to these insects, or the petiole contain more sap. Occasionally tiny flies and parasitic wasps are attracted to the dense aggregates of aphids on cacao flowers. Leafhoppers, shield bugs, grasshoppers, and many other insects thrive in cacao trees. Yet the insects and other creatures that dwell in cacao on plantations represent only a small fraction of the species found in the tropical rain forest whence they came.

Squirrels and monkeys chew large, gaping holes in the tree's ripe pods and set the stage for the cycle of rot and decay that eventually provides an ideal breeding niche for some of the tiny midges that pollinate the tree's flowers. Thus the cacao tree, in exchange for this pod damage, breeds its own pollinators, needed to make future seeds.

It is easy to understand why animals would seek out these pods in the wild. I have quenched my thirst many times by sucking on the sweet seed pulp exposed by a machete's sharp steel. The pods, by their size, shape, color and texture, are well adapted to seed dispersal by monkeys and other animals capable of chewing all the way through the tough pod wall. In wild cacao, most of the pods would be close to the canopy branches of the trees, readily accessible to monkeys and parrots. Undoubtedly, the pod colors, especially the rich yellows and oranges, attract these visually oriented animals (Janson 1983).

Monkeys are the chief seed dispersers of *Theobroma* and *Herrania* in Amazonian rain forests (Janson 1983). When a monkey extracts the slippery insides from one of the pods, it carries off the precious bundle, like a bunch of grapes, to some other place in the forest canopy, where it sucks the pulp from the almond-sized seeds. The bitter-tasting seeds get dropped to the forest floor and become part of the rain forest's collective future. The seeds become extra slick from the animal's saliva, along with a thin coating of remaining pulp mucilage. This slipperiness probably contributes to the likelihood that the seeds are dropped by the animal without severe damage.

The monkey apparently is not after the bitter-flavored cacao seeds. The seed, through the bitter taste endowed upon it by the high concentrations of alkaloids in the cotyledons, most notably caffeine and theobromine, protects itself from being eaten while being handled by an animal. The cotyledons, rich in fats and oils (food for the embryo), otherwise would be very attractive targets as food for monkeys and other animals.

There is a very strong connection between the cacao seed dispersers and the cacao pollinators in the tropical rain forest. Monkey- or squirrel-damaged pods rot out and provide an ideal breeding site for some species of cacao-pollinating midges. The survival of *T. cacao* and undoubtedly other species of *Theobroma*, as well as *Herrania*, in their natural forest habitats hinges in large measure upon the interactions of these trees with flies and monkeys. The cacao pod does not fall off the tree when ripe. Rather, the tree closely depends upon monkeys and other sharp-eyed mammals to discover the pods in the muted stippled light of the forest subcanopy and break them open to release the seeds.

It is not known whether cacao secondarily lost the ability to drop or shed its pods, or if the tree never evolved this habit in the first place. Dropping mature pods certainly could be adaptive in the sense that it removes reservoirs of pathogenic microorganisms that would otherwise be close to and might infect new, developing pods. Yet the adaptive value of pod retention must have something to do with the clear, close dependency upon arboreal animals as the primary seed dispersers for cacao in the tropical rain forest.

Furthermore, the soft, pulp-covered seeds of cacao, once outside of the pod, tend to mold and rot quickly. If ripe pods dropped off the tree still covered with pulp, the seeds would likely rot before they had a chance to germinate. Also, undesirable crowding effects beneath pod-shedding cacao trees could result in further losses through competition among new seedlings. Therefore, staying on the tree, encased in the pathogen-free internal environment of the pods and waiting to be released and dispersed by animals to distant corners of the rain forest is likely a most prudent reproductive strategy.

The fact that many rain forest tree species drop their mature fruit to the ground where the seeds are then dispersed by animals such as rodents (Janzen 1971; Smythe 1970) implies that there could be intense competition among tree species for what must be a limited supply of seed-dispersing, ground-level animals. Therefore, a selective premium can be further placed on fruit retention, with seed dispersal mediated by arboreal animals, as a means of slackening this competition and ensuring the survival of the seeds. Other species of *Theobroma* whose pods drop off the tree include species such as *T. bicolor* that inhabit the rocky edges of rivers, where the thin-walled pods (Cuatrecasas 1964) unique to these species are easily shattered when they fall, releasing the seeds to be swept away in the water.

Breeding Strategy

Wild cacao trees often grow as widely scattered clumps in the rain forest, the result, perhaps, of monkeys dropping several seeds at a time. A clump of wild cacao trees might also arise from vegetative growth of new shoots from an original tree trunk. The interplay of these two mechanisms of clumping, along with the generally low number of flowers in contrast to the many flowers of cultivated cacao and the low number of pods, suggests that wild cacao originally evolved with what many biologists would call a conservative breeding strategy.

A species with a conservative and specialized breeding strategy allocates a relatively low amount of energy or resources to the production of flowers, seeds, and fruit. Relatively more energy is channeled into vegetative processes. Coupled with this would be a long adult life span and a well-developed system of defenses against predators and pathogens, which might be the function of alkaloids and other pesticidal substances in the leaves, flowers, wood, and seeds of cacao. Cacao's mycorrhizal fungal associations suggest yet another component of a specialized adaptive strategy.

The large seeds of cacao contain enough stored energy to allow a new seedling to grow for several weeks before it requires ambient foodstuffs such as photosynthates. The seeds germinate very quickly once removed from the pods. Collectively these biological traits, and the low levels of natural pollination, endow wild cacao with a typically conservative reproductive proclivity, one in which asexual propagation of discrete genotypes, each more or less adapted to a particular set of ecological conditions in the rain forest, is held at a selective or evolutionary premium over the crossing of genetic material through sex. This is true even though the flowers, with their hermaphroditic state, would seem to facilitate sexual behavior. The fact that the flowers are structurally designed to minimize self-fertilization, however, lends further credence to the portrayal of wild cacao as a sexually conservative species of the tropical rain forest's substrate.

Various other species of *Theobroma*, along with the allied *Herrania*, which have not been cultivated in the manner of *T. cacao*, reflect in large measure the conservative reproductive condition of the tropical forests in which they evolved long ago. Species of *Theobroma*, under the most widely accepted classification (Cuatrecasas 1964), are distinguished from one another not only by flower morphology but also by branching patterns of trees and mode of seed germination. *Theobroma* thrives best within 18 degrees north and 15 degrees south of the equator under conditions of heavy rainfall and high temperatures throughout the year, and with the high humidity and dense shade of the rain forest. *Theobromas* generally are not trees of open spaces. They evolved under closed-canopy circumstances, and much of the little that is known of their natural history reflects this kind of evolutionary history. Conditions of dense forest shade tend to suppress or dampen the level of flowering in trees, and this is certainly what is found in forest populations of *Theobromas*, including wild cacao. Low-level flowering under natural conditions suggests a conservative breeding strategy in rain forest tree species within the lower strata of the forest.

Although there have been very few studies of wild *T. cacao* in its native habitat, recent observation by John Allen (1982) provides a helpful starting point. Allen was able to collect many variants of *cacao silvestre*, or *cacao de monte* (wild *T. cacao* from the Río Napo region of Ecuador on the eastern slopes of the Andes, the probable origin of cacao). Seeds were obtained from trees up to 20 meters tall and having two to five thick, leaning trunks. Collecting in the Napo region of Ecuador on the eastern slopes of the Andes, Allen discovered a spatial distribution of about five cacao trees per hectare of rain forest. Trees within the forest possessed sparse crowns of leaves that were limited to the upper reaches of the branches, whereas wild cacao trees in open sites along rivers had denser, more symmetrical canopies or crowns down to within a few meters of the ground.

Wild cacao trees, which may grow to 20 meters or so in height—typically more than double that of cultivated cacao trees—develop multiple, leaning trunks. Old trunks that fall over often sprout new trees, and multiple new trunks frequently sprout from the base of an old trunk. Even though old trunks may eventually die, the tree's genetic constitution is immortalized through this vegetative propagation. Wild cacao trees channel much of their energy into growing new trees from a primary or old trunk, and less into sexual breeding of new trees. The trees bear few flowers and even fewer pods.

Allen also noted that the generally reddish flowers were confined to the upper portions of these wild trees, and the pronounced cauliflory of the plantation trees was not present. Unlike the flushing of fresh leaves in plantation cacao trees, wild cacao leaf flushes were pale green and not red. Wild cacao trees, products of the natural environment, thus appear very different in growth-form and breeding habits from the carefully pruned trees in plantations.

Allen found different variants occurring in different areas of the Amazon Basin, suggesting considerable localized genetic differentiation and adaptations to subtle features of the environment. Under the relatively predictable environmental conditions of the forest shade, small, scattered pockets of wild cacao trees may more or less independently adapt to selective pressures operating on a microscale. This would circumvent the usual high selective premium placed upon outcrossing mediated through sexual reproduction—nature's trump card for endowing species with the evolutionary resiliency often needed to cope with unpredictably changing environments. Wild cacao trees, therefore, might have little evolutionary advantage in producing lots of flowers, high levels of natural pollination, and lots of seeds. In fact, just the opposite seems to be the natural state of the cacao tree. And this condition, of course, becomes the dilemma of the cacao farmer.

Breeding and Self-Incompatibility

A species such as cacao, which evolved within the predictable environment of the tropical rain forest (Slobodkin and Sanders 1969) and possesses high longevity, low reproductive output, stringent chemical defenses, and specialist associations with other species, appears to have traits opposite of what would be desirable in a perennial cash-crop tree. Yet the discovery and quest for chocolate has prompted humankind over thousands of years to coax this species out of the rain forest and into cultivation.

Outside of its natural habitat cacao appears to be a very different creature. This is an artifact of tinkering by humankind, which has trimmed and groomed the tree, much like a pedigreed dog or a variety of garden rose, to its liking—in this case to become a manageable organism capable of producing chocolate. The challenge of doing so has been magnified by a serious lack of understanding about the genetics of cacao that underlie the range of variation observed both in nature and through programs of artificial breeding and selection.

The basic design of most cacao plantations in Central America, in which many trees are planted in rows and blocks, is a very different spatial arrangement from the small clumps of trees scattered over great distances in the rain forest. The tree is compelled to become liberally reproductive in the plantation to satisfy an agronomic mission, as opposed to the species' highly specialized conservative ecological habits in its natural state.

What is also readily evident in the cacao plantation is the presence of different varieties of the species, those rendered by breeding programs of artificial selection. Many trees are different from others not only in the color of the mature pods, ranging from green and yellow to orange, bright red, and maroon, but in the positioning of the flowers and their sizes. In some varieties, the flowers are arranged in large clumps on the trunk, with as many as a hundred floral buds, while in others the clumps are much smaller. Trees with large clumps of flowers are usually genetically self-incompatible: Pollen from one flower cannot effectively pollinate the same flower or other flowers on the same tree. Flowers on such trees tend to be more pinkish or reddish, although this is not universally true. Other trees, in which flower densities are generally much lower, are typically self-compatible, capable of effective self-pollination.

A. F. Posnette wrote five decades ago that Amazonian varieties of cacao are basically self-incompatible, thereby requiring cross-pollination among different trees for the flowers to set fruit (Posnette 1944, 1945). Although each flower possesses both male and female reproductive organs, which would theoretically allow for self-pollination among the flowers on a single tree, and within each flower, the pollen grains from one flower cannot successfully fertilize the egg cells (ovules) of the same flower. In fact, an incompatibility factor impedes the pollen's journey before it even has a chance to fertilize the egg cells. The pollen tubes, which transport the male sex cells down the flower's style into the ovary, cease to grow properly in cases of incompatibility and are stymied within the ovary tissues just short of the egg cells. Cultivated cacao, however, exists in both self-compatible and self-incompatible forms. It is also possible that self-compatible forms exist in the wild but have not been discovered. No one knows for sure the breeding system of truly wild cacao. It is also very difficult to locate and discern a veritable wild population of this species.

At the crux of the evolution of breeding systems is the high adaptive value associated with genetic outcrossing within the breeding population in relation to the mode of pollination and the kinds of animals that function as pollinators in many species (Bawa and Beach 1981). Outcrossing, or cross-pollination, is highly adaptive in that it promotes genetic heterogeneity, a basis for sustaining some level of flexibility in coping with environmental changes. In some tropical plants, only male flowers are produced on an individual tree, while female flowers occur on a second tree, a condition called dioecy. Obviously these plants require cross-pollination. In other species, such as cacao, flowers are morphologically and physiologically hermaphroditic, containing both male and female sexual organs (monoecy). If the tree is genetically self-compatible, then self-pollination can result in fruit-set. If the tree is self-incompatible, then pollen must move from that tree to others of the same species for fruit to be set.

One of the great mysteries about the origin of cacao centers on its genetic compatibility or lack thereof. Certainly from an agronomic standpoint, the maintenance of genetic variation through the existence of self-incompatible trees, which require cross-pollination to set fruit, would be desirable as a means for certain varieties or clones to combat certain kinds of diseases. Genetically incompatible cacao would also require the availability of strong, long-distance pollinators to transport pollen from one population to another. Cacao flowers within a clump of self-compatible trees are effectively pollinated by short-distance pollinators.

The incompatibility phenomenon in cacao flowers was first reported by F. J. Pound in Trinidad (Pound 1932a). Pound's discoveries came from controlled studies in which flowers were hand-pollinated with pollen from different cultivars or varieties. He observed that certain cacao trees could not set fruit with their own pollen, while others could. Self-incompatible trees often turned out to be good cross-pollinators, even though a high proportion of these require pollen from self-compatible trees for successful pollination. The site of the compatibility mechanism in cacao, unlike most other plant species, in which it is on the style or stigma, is unique in that it occurs at the point of fusion between the sex cells (gametes) in the ovary.

Interestingly, Amazon-derived cultivars of cacao are all self-incompatible, suggesting that there might have been strong selection in nature for outcrossing in wild populations of *T. cacao*. Self-incompatibility may have been the ancestral genetic condition of wild cacao before the tree came into cultivation. In its wild state on the lower Andean slopes of Amazonia, cacao occurs as scattered clumps of tall trees along streams and other openings in rain forest (Allen 1982). If wild cacao is self-incompatible, pollinators capable of long-distance flight between clumps of these trees would be required for sexual reproduction. Such an arrangement would ensure the maintenance of some level of genetic variation.

An important implication of cacao's self-incompatibility has to do with this species' cultivation as a plantation crop in the lowland humid or wet tropics. Typically plantations—whose mission is to produce either cocoa beans for chocolate consumption, propagation material in the form of seeds or budding stocks, or both—contain mixtures of self-compatible and self-incompatible trees. Cacao growers must often plant many different varieties of trees in their plantations as a way of ensuring that there will be adequate cross-pollination for self-incompatible trees to set fruit. Although they may have lower levels of fruit-set, self-incompatible trees are generally advantageous to cacao growers in that they tend to be more disease resistant and have other desirable agronomic properties. Establishing the nature of the incompatibility mechanisms in cacao has been a major challenge in establishing agronomic practices aimed at enhancing production of cacao beans in the commercial state (Cope 1939, 1958).

Studies over the past several decades have focused upon understanding the inheritance of the incompatibility factor in cacao. There appears to be a series of "S alleles" (an allele is an alternate state of a gene on a chromosome) with a definitive order of genetic dominance and that regulates the incompatibility of the plant and its ability to cross-breed. The results of cacao breeding studies on the possible occurrence of hybrid vigor have been obscured by a lack of knowledge about the basic genetics of the species. Therefore, attempts to produce successful hybrid seed in cacao have met with highly mixed and unpredictable results. Thus while it may be desirable to produce viable hybrid seed in cacao as a means of improving yields, a lack of understanding of this tree's genetics has made it very difficult to produce genuine hybrid seed.

Genetics and Diversity

The spectrum of variability among the trees, aside from age and micro-ecological effects, is a function of each one's genetic background. Genetic differences among varieties, even among individuals of a single variety, influence hardiness or resistance to diseases, flowering capacity, fruit-load tolerance, foliage growth, and so forth. It can also be a matter of how different varieties vary in their attractiveness to pollinating insects (Chapter 5). Moving the cacao tree, as a species, from the rain forest into the cacao *finca* may also change the tree's relationship with soil biota, including symbiotic species of fungi, which in turn affects the pollination by midges.

From what little is known about cacao genetics, and from field observations on variable characters of the pods, seeds, flowers and leaves of the tree, there exists considerable variability (presumably genetic) among different cultivars for determinants of yield. All of these elements seemed to be controlled by single "major" genes, meaning that each variable is a single-gene trait and modifier genes affecting the major genes (Soria 1978). The size, weight, and production of pods and seeds are controlled by quantitatively inherited genetic factors.

What agronomists have come to appreciate is the apparent high degree of genetic variability in cacao that has been discovered in wild populations in the Amazon and Orinoco river systems (Chapter 1), providing a source of new varieties. Yet it is not enough to merely collect and identify new varieties of wild cacao and hope that they will be useful in developing new "hybrid" forms that perform in a superior agronomic manner. New varieties must be evaluated for their desirable attributes and the manner in which these stand up under breeding programs. It may take a decade or longer to prove disease resistance and its heritability. Nurseries of newly acquired cacao varieties are being established for this purpose and to maintain these varieties successfully over long periods as sources of material for commercial propagation and selective breeding with other varieties to improve crop performances. For example, some of the material collected in the 1980s by John Allen in Ecuadorian rain forests may provide genes with enhanced resistance to the dreaded witches-broom disease, which annually wipes out cacao harvests in South America.

From Allen's nursery at Napo in Ecuador, propagated material is transferred to other key "gene bank" centers for cacao in other parts of the world to augment existing cultivar collections or establish new ones. Plant breeders can then request "budwood" from desirable cultivars. These short lengths of young branches bearing dormant growth buds can then be grafted ("budded") onto the stems of cacao seedlings to be propagated or used in breeding experiments. Successful grafting results in new apical growth on the seedling exhibiting the genetic characteristics of the grafted budwood. Although such grafting is tedious, with often low rates of grafts taking hold, the propagation of cacao trees with budwood is a faster means of obtaining desired varieties than having to rely upon seeds and crosses.

Growing a cacao tree from seed requires much more time than grafting one tree stock onto another. Because cultivated cacao trees are highly susceptible to many kinds of disease and pests, partly a result of being grown in large stands and partly because of an apparently low amount of genetic diversity in some cultivars, the long-term future of chocolate commerce depends, in large measure, upon the preservation of tropical rain forests in South America, and to some degree, in Central America. These forests are the primary source for new, spontaneously arising cacao varieties, some of which may possess agronomically desirable traits. Without sources of wild cacao, the only sources of genetic material would be the gene banks already established, and these may not be sufficient in the long run. New forms continually arise in natural populations of species, even though, in the case of cacao, this process of sexual selection may require several years between generations. The success of cacao as a commercial crop is therefore closely linked with the preservation of tropical nature.

Nature in the Cacao

Mysteries of Pollination

No matter where cacao is grown in the world, the trees set very few pods in spite of producing prodigious numbers of flowers. This paradox was first brought to my attention by Bob Hunter at his farm, La Tirimbina, in Sarapiquí, Costa Rica. It appeared as if pollination by natural agents was a prominent limiting factor in overall production wherever cacao was cultivated. This had been a fairly widespread perception among people studying cacao pollination, although not without controversy. Were pollinating insects scarce in cacao plantations? If so, could there be a means of encouraging their proliferation in these habitats? These were the kinds of questions that began to emerge as I read the available literature on cacao pollination and began developing my own ideas and directions for possible research.

It was logical for me to begin my studies of cacao pollination at La Tirimbina, since I had already spent many years there doing other kinds of research with insects (Young 1991). La Tirimbina sits next to two other *fincas*, El Uno and La Tigra, where considerable tracts of cacao had been in cultivation for many years. The cacao had been planted in two distinctive arrangements on the two farms, providing me with an ideal situation to discover how populations of cacao pollinators differed in the two types of habitats.

In El Uno, the cacao grew under a shade cover of Pará rubber trees. In La Tigra, the cacao had been planted under the remaining canopy-size trees of a thinned-out mixed primary and secondary rain forest, containing many tree species. Some of the La Tigra cacao was mixed with planted timber trees like laurel *(Cordia alliodora)* and manu *(Minquartia guinanensis)*. But most of the overstory in La Tigra was, and is, natural vegetation, chiefly older secondary rain forest. I was intrigued, early on, by the possible effects of a highly diversified ground layer of leaf litter (La Tigra) compared with those of a simple polyculture farm (El Uno) on both the diversity and abundance of cacao pollinators. I suspected that the leaf litter beneath an overstory of natural vegetation in a cacao plantation supported a richer assemblage of pollinators than did a uniform shade cover of Pará rubber. This idea became an early focus of field experiments in the Sarapiquí farm.

One of the first things I learned was how cacao flowers could be hand-pollinated to either make specific hybrid crosses or to supplement the poor levels of natural pollination. Hand-pollination yielded greater numbers of mature pods than did natural pollination. Thus it certainly appeared that cultivated cacao is pollinator-limited—that is, natural pollination is a limiting factor in the production of pods.

If this assertion were true, then a scarcity of pollinators in commercial plantations might be behind the typically observed low pod yields. Other researchers in different cacao-growing regions had observed that newly planted cacao, following clearing of some, but not all, rain forest had high pollination levels. Could it be, therefore, that established cacao plantations were no longer ecologically suitable for sustaining populations of pollinators?

Midges and Other Pollinators

I started my research under the assumption that tiny flies of the dipteran family Ceratopogonidae, in particular, various species of *Euprojoannisia* and *Forcipomyia*, are the chief natural pollinators of cacao (Bystrak and Wirth 1978), as most research papers indicated (e.g., Entwistle 1972). The ceratopogonids, commonly referred to as "biting midges" because they include the pesky no-see-ums, are distributed worldwide. Cecidomyiids, or gall midges, were also suspected to be cacao pollinators (e.g., Kaufmann 1973).

I also discovered from the literature that most examples of dipteran-pollinated flowers portray the flies as occasional generalist pollinators, having little or no specialized ecological dependency on the species of plants in question, and vice versa (e.g., Percival 1965; Proctor 1978). As adults, the midges undoubtedly feed on many things, probably including pollen grains and nectar from various species of plants. Therefore, although the specialized structure of the cacao flower suggests specialization for a certain kind of pollinator, the habits of these particular pollinators are presumably not as restrictive; these midges are not highly dependent upon cacao flowers for their survival. In the Amazonian rain forest where *Theobroma* evolved, these ecological opportunists undoubtedly foraged for pollen from the flowers of many plant species, and one would therefore not expect a high degree of fidelity between cacao flowers and pollinating midges in plantation settings.

Up until 1977, as a field biologist in Sarapiquí, I had been studying large-bodied insects such as cicadas and butterflies (Young 1991). The switch to midges, resembling little more than flying specks of dust, was quite a logistical challenge. Most pollination biologists work with large-bodied, relatively easy-to-observe animals such as bees, hawkmoths, bats, and hummingbirds (e.g., Feinsinger 1983; Kevan and Baker 1983). Such organisms are also relatively easy to mark for studying their movement patterns, site fidelity, and much more. But midges? This was a whole different story—one filled with lots of challenges and frustrations. Although there have been fascinating discoveries of flies as the principal pollinators in some lilies such as *Arum* (e.g., Meeuse 1975), by and large, the study of fly-pollination systems in tropical plants has been thin.

Moreover, the role of Diptera as pollinators has always been somewhat controversial. One traditional view holds flies to be unspecialized, opportunistic, and only occasional pollinators, lacking the more specialized energy-demanding adaptations of bees, sphingid moths, hummingbirds, and bats—groups especially significant in the highly diversified pollination schemes of the New World tropics. Yet, even from pollination studies in the temperate zone, there is some compelling evidence for specialized interactions between flies and flowers. Fungus gnats (Sciaridae and Mycetophilidae), for example, are important pollinators of certain kinds of orchids in northern California (Mesler et al. 1980), resulting in high rates of fruit-set. Fungus gnats are also pollinators in some Araceae, where floral fragrance mimics the odors of fungi, the breeding substrate for these insects (Vogel 1978).

Insect-pollinated species represent a major reproductive pattern in plant communities, especially in the tropics (Proctor 1978). Under these conditions, closely related plant species, of which there are often many in tropical areas, may have different sets of pollinating insects, an effective means of preventing the formation of intersterile hybrids and therefore ensuring reproductive success in the wild. That *Theobroma* flowers are visited chiefly by biting midges and gall midges (e.g., Posnette 1944; Billes 1941; Cope 1939; Glendinning 1962; S. de J. Soria 1970, 1977a, 1977b; Soria and Wirth 1979; Hernández 1965; Wirth and Waugh 1976; Young 1985a, 1985b, 1985c, 1985d), and those of the closely related *Herrania* (Schultes 1958) by phorid flies, provides compelling evidence for such an adaptation, partitioning pollinator types between closely related genera of plants (Posnette 1944; Young 1984b).

When I started studying cacao pollination I did not rule out bees as pollinators of cacao. There had been scattered, occasional reports of bees visiting cacao flowers (e.g., Walker 1959; Kaufmann 1975; Soria 1975). One of the first things I did in the La Tigra cacao was to investigate the behavior of *Trigona* (stingless bees in the apid subfamily Meliponinae) on cacao flowers. It was commonly observed that these bees visited cacao flowers, especially those bathed in sunlight (Young 1985b). At La Tigra the exquisite little amber-colored *Trigona jaty* was known to frequently visit cacao flowers, suggesting a possible solution to the problem of low natural pollination. Managing colonies of such a social bee would have been a cleaner, more practical (from an agronomic standpoint) solution to the problem of low pollination levels in *T. cacao* than dealing with midges. Artificial nesting boxes or bee containers would have been relatively easy to devise and distribute in cacao plantations.

I had read one previously published account of *Trigona jaty* visiting cacao flowers in Costa Rica, observations made at La Lola in the 1960s as part of a doctoral thesis (Hernández 1965). Another account reported *T. jaty* visiting cacao flowers in Bahia, Brazil (Soria 1975). Watching these bees at La Tigra, I was inspired to find out if this 3.5-millimeter bee, an ideal "fit" for a cacao flower, was indeed a pollinator both here and at La Lola.

My approach was to place a colony of *T. jaty* inside a large cage enclosing two cacao trees. A second identical cage without bees and enclosing two other cacao trees would serve as a control. The two wooden-frame cages were each 6 meters long by 3 meters wide by 3 meters tall and equipped with a little door, which I had to crawl through to observe bee behavior and count flowers. A local farm worker had given me a healthy hive of *T. jaty* that he had transferred months before from a tree hole cavity into an empty "monkey pot" fruit husk (*Lecythis costaricensis*). A piece of corrugated metal roofing, propped up with wooden stakes, protected the hive from rain. A plastic screen mesh stretched over the wooden frames allowed tiny flying insects, such as midges, to pass through, but not bees. Any increase in pod-set in the experimental cage, coupled with observation of bees, could therefore be attributed to the bees.

Colony of the stingless bee *Trigona jaty* Smith, housed in the husk of the monkey pot tree fruit and suspended inside an experimental cage enclosing two flowering cacao trees. Inside the tree-covering cage, the bee nest is suspended beneath a small, rooflike structure to protect it from the rain. The bee nest was used to determine if *T. jaty* could successfully cross-pollinate cacao flowers on the two trees enclosed in the cage. Several bees can be spotted in the lighted airspace just to the left of the vertical stake holding the nest.

My goal was to determine whether *T. jaty* visited cacao flowers but did not pollinate them. I could tell from just watching the movements of *T. jaty* on cacao flowers that there was little chance they pollinated them, but perhaps I was not seeing some indirect effect of the bees—such as a "buzz" pollination, general jostling, or some other less definable effect that could result in pollination. Buzz pollination is the process by which bees vibrate their bodies at high frequency in contact with anthers, causing the rapid release of pollen, which gets deposited on the stigma of that flower or moved to other flowers. I strongly suspected that *Theobroma* flowers, especially those of *T. cacao*, did not exemplify this pollination syndrome. Buzz pollination seemed unlikely, given the viscous nature of cacao pollen, which makes it difficult to dislodge from flowers. My null hypothesis was that there would be no significant differences in the numbers of new pods set on cacao trees in the two cages following a specified period of time.

For several weeks after the bee hive was removed at the end of the study, the four trees in the cages were examined for the presence of newly formed pods. I was unable to reject the null hypothesis regarding pod-set differences between the trees in the control and experimental cages. The lack of significant increases in the numbers of new pods in the experimental cage, together with what others had observed in the field, certainly reinforced the strong likelihood that *T. jaty* was an effective cacao "pollen thief" rather than pollinator (Young 1981). A pollen thief is a floral visitor that steals pollen or nectar without effectively pollinating the flowers.

The experiment sometimes brought me face to face with creatures besides midges and bees. The cages seemed to be inviting to other creatures, including many large golden orb spiders, whose dense webs I had to keep knocking down lest they trap my bees. On one visit to the control cage, I encountered an even more unwelcome guest, a one-and-a-half-meter long fer-de-lance, *Bothrops asper*, nestled in the leaf litter. When I crawled into the cage, I had hardly noticed the deadly viper. Its splendid mottled earth tones and that (appropriately) rich chocolate brown velvet-sheened wedge of color on top of its massive head gave the snake a near-perfect camouflage with the dead leaves. Although it was stretched out along one edge of the cage, not coiled and ready to strike, upon noticing it, I slipped back ever so slowly and as quietly as possible gently closed the door to the cage. I would wait a week or two to recheck the control cage, giving the snake wide berth and hoping that it would move on soon, which it did eventually.

My intrigue with bees as potential pollinators of cacao did not end with these

The original building of CATIE in Turrialba, Cartago Province, Costa Rica, as it appeared in 1986, and environs.

View of a typical cacao plantation in Central America.

Cauliflory, the growth of flowers directly from the trunk, on a cacao tree.

Trigona bee visiting a cacao flower.

Worker splitting open ripe cacao pods to remove seeds, which are collected in sturdy wooden boxes, as shown. The husks are left in piles, where they rot, collect rainwater, and become breeding sites for a variety of insects, including potential pollinators.

A ripe cacao pod split open crosswise to show the thick pod wall and the arrangement of seeds inside. The almond-sized, dark reddish-brown seeds are encased in a slippery white pulp.

Ripening pods on the trunk of cacao close relative *Herrania*, a striking example of cauliflory in the "wild cacaos."

Flushing of new red leaves on cacao trees in a plantation.

Ripening pods on cacao trees in a plantation, and in a pile after harvesting.

Collecting seeds from newly harvested cacao pods.

Drying and sorting cacao seeds in the sun at the La Lola Experimental Farm near Siquirres, Limón Province, Costa Rica. The worker runs his hands through the drying beans to turn them.

Drying cacao in the Dominican Republic.

cage studies. Several years later, I became involved in a collaborative field study of honey bees with Eric H. Erickson, Jr., a honey bee specialist, and his wife, Barbara, an insect physiologist, both affiliated with the University of Wisconsin, Madison. This week-long project was to determine whether or not Africanized honey bees (*Apis mellifera scutellata*) were visiting cacao flowers, that is, collecting pollen from them. If honey bees could be cacao pollinators, especially tropically adapted forms such as the Africanized honey bee, pollination of the crop might be easily managed.

With the help of William Ramírez of the University of Costa Rica, we obtained eight colonies of Africanized honey bees, transported by pickup truck from San José to La Lola, three and a half hours across the mountains. Ramírez had been studying Africanized honey bees since their arrival in Costa Rica in 1982 or 1983. Marla Spivak, a graduate student working with Orley (Chip) Taylorb in Ramírez's laboratory, accompanied us to La Lola and assisted with the study.

We introduced the colonies into the La Lola cacao for six days in September, forewarning the farm workers to stay clear of the colonies for this period. Four of the colonies were set up along one border of the cacao forest (a strip of abandoned cacao) and the other four in a more central location in the La Lola cacao. Each colony was outfitted with a special "pollen trap" that dislodged pollen pellets from the bees' legs as they reentered the hive after foraging. By collecting the pollen pellets at the end of each day for several days, and then freezing and analyzing the samples, it was possible to determine the kinds of plant species visited by the bees, especially at the family and generic levels. Each kind of plant produces a unique pollen grain, a sort of biological fingerprint.

Aliquots of the resulting forty-eight frozen samples of pollen pellets were sorted back in Madison on the basis of the texture and color of pollen grains under a light microscope. A subset of seventeen distinctive pollen types was then sent to a pollen-identification company in Pennsylvania for taxonomic determinations. These subsamples revealed eleven major pollen types common to the hives in both areas of the cacao (Erickson et al. 1988). Pollen from corn, other grasses, and fig trees made up about 89 percent of our samples from the central colonies, while fig pollen made up 89 percent of the pollen found in bee colonies at the border of the La Lola cacao. We found no cacao pollen at all in the samples, and no honey bees were seen visiting cacao flowers. These observations confirmed that honey bees, even the Africanized bee, do not visit or pollinate cacao flowers (Erickson et al. 1988).

Pollinator Breeding Site Studies

The bee studies were sidelines to my major research on the breeding sites of biting midges, the principal pollinators of cacao. My initial research proposal to the American Cocoa Research Institute was to spread predetermined plots within the El Uno and La Tigra cacao groves with various kinds of rotting plant debris, to answer three questions. First, do biting midges prefer one kind of breeding substrate over others? Second, is there a marked improvement in the level of fruit-set in cacao when these treatments are applied in plantations? Third, are there any differences in the effectiveness of breeding substrates in building up midge populations and enhancing fruit-set within a densely shaded cacao and rubber tree habitat (El Uno) versus a less shaded, thinned-out rain forest planted with cacao beneath the canopy (La Tigra)?

There had been several reports in the published literature about the use of plant debris as breeding sites for biting midges. Some species of cacao-pollinating ceratopogonid midges were found to breed in the matted debris that accumulates in the leaf axils of tank bromeliads in Costa Rica (Privat 1979). Some studies in Brazil, especially those of Saulo de Jesus Soria and Australian biologist John Winder (e.g., Winder 1972, 1977), revealed that various species of midges breed in bromeliads and many other kinds of plant debris, including rotten banana pseudostems, cacao pod husks, and leaf litter. This information had been gathered chiefly by going out into plantations and sampling these substrates for the fairly distinctive-looking larvae and pupae of the tiny midges. But apparently no one had experimented with plots of test substrates in cacao plantations and documented the subsequent "colonization" of them by midge species. More than twenty years earlier, however, a paper published by a French entomologist based on field studies in Africa suggested that midge populations could be increased in cacao plantations by adding organic trash in which these insects could breed (Dessart 1961). Such studies prompted me to set up field trials on midge breeding sites at the Sarapiquí *fincas* and, later, at La Lola.

The Sarapiquí study was done in three time segments—June and July 1978, February and March 1979, and August 1979—to complete one approximately twelve-month cycle of two rainy seasons, each occurring on either side of one tropical dry season. Replicated piles of sliced *Musa* pseudostems, made by cutting up the trunks of fallen banana plants with a machete, were placed on the ground beneath cacao trees at each site, as were piles of freshly opened pod husks around other trees. "Bromeliads" were fashioned from plastic drinking cups outfitted with whorls of plastic strips, spray-painted green, and perforated on the bottoms to allow drainage of excess rainwater.

Chunks of sliced-up banana pseudostems in the leaf litter on the cacao plantation. These pieces are in an advanced stage of natural decay, when they attract cacao-associated ceratopogonid midges, which breed and undergo their life cycles in the rotting plant matter. Many other arthropods also thrive in these chunks of rotted pseudostems, which remain fairly moist well into the tropical dry season, while the surrounding leaf litter dries out. Piles of these pseudostem pieces are placed beneath many cacao trees under the assumption that increased breeding by midges in this debris will produce a larger number of adult midges to pollinate nearby cacao flowers.

Earlier, I had written to Richard C. Lewontin, the distinguished population geneticist at the Museum of Comparative Zoology at Harvard University and my former population genetics professor at the University of Chicago, regarding the possible use of artificial bromeliads to examine the colonization of arboreal insect communities or assemblages in water-filled containers. Somewhere, many years before, perhaps even while I was a graduate student and Lewontin was on the zoology faculty at the University of Chicago, I had heard that the late Robert MacArthur, the eminent Princeton University ecologist, had thought about using artificial bromeliads to examine patterns of species richness in tropical rain forests. Lewontin wrote back confirming my recollection of MacArthur's project, supporting the potential of using artificial bromeliads for this purpose.

I later obtained larger and sturdier plastic cups, free of charge from the Sweetheart Cup Corporation in Chicago, to devise a different model of bromeliad. The "new" bromeliads were unpainted but perforated and attached by sturdy guide wires to the branches of cacao trees in both study areas. This model would prove to be more sturdy than the earlier, albeit more artful, version. The "bromeliads" were filled with rotting leaf litter, twigs, mosses, and other debris. Two-by-two-meter-square wooden frames were placed on the ground beneath cacao trees and filled with either cacao leaf litter or a mixture of leaf litter and rotted pod husks. Control plots in each habitat were two-by-two-square-meter areas of natural leaf litter. Experimental leaf litter plots were much deeper, representing about a four-fold increase in mulch.

My various treatment substrates were distributed in each plantation using a random-numbers table and by matching treatments and replicates with number-coded, previously tagged cacao trees. I had been gathering monthly phenological data on quantities of flowers and pods on many tagged trees in El Uno and La Tigra, and this census program continued through my insect population studies. Keeping track of flowers and pods would enable me to detect a possible relationship between the locations of substrate treatments, their resident midge populations, and the abundance of new pods (a measure of fruit-set) on individually tagged trees. I also used these data sets, taken repeatedly here and subsequently at other localities in Costa Rica, to describe seasonal patterns of flowering and fruit-set in cacao in Costa Rica (Young 1984c).

An artificial bromeliad suspended from the branch of a cacao tree on the La Tigra plantation in Sarapiquí. Leaf litter is added to the plastic cup, which is perforated at the bottom to drain excess rainwater. The litter in the cup is dumped out and examined periodically to determine if cacao-associated midges are breeding there. Some of these midges are known to breed in rotting debris found in real bromeliads.

Wooden frames are filled with leaf litter and set around the bases of mature cacao trees to increase the availability of breeding sites for cacao-associated midges. Bags of leaf litter are gathered up periodically from these frames and examined closely for the larvae and pupal stages of midges.

For this initial year-long study, the various test substrates were replenished several times, an especially labor-intensive chore when it came to preparing the banana pseudostem slices. Farm workers dumped large piles of the cut stem slices for my studies, getting them from nearby stands of bananas along a path both at El Uno and La Tigra, and I used a large gunnysack to haul all the sliced-up stems to the individual plots. Here, and later at La Lola, I hired local students and plantation workers to assist with replenishing the breeding substrates and recording basic phenological data from tagged trees in the study plots.

The basis for using banana pseudostems and other organic mulch is the assumption that mulch-breeding midges are greatly limited in their capacity to proliferate and spread through the cacao plantation because these "preferred" types of mulch were usually wanting. Plantation workers are often instructed to clear away rotted pod husks and other natural debris as a sanitary measure against the spread of pathogenic fungi capable of attacking growing cacao pods. I reasoned that adding more preferred mulch might induce the midges to "colonize" the plantation from the nearby rain forest and build up larger populations in response to the plentiful breeding sites. This approach simply played upon the basic ecological principle that populations of animals grow in response to the availability of adequate resources required for breeding. But its application to cacao-pollinating midge populations was new.

The use of machete-cut slices of banana pseudostem, called *bastago* by the Costa Rican plantation workers, was especially appealing because of its dense, compartmentalized structure, ideal for breeding midges. Left to rot after being cut, the *bastago* in the leaf litter beneath the cacao trees maintains its form and structure even though soft tissues rot quickly. As it rots, this pseudostem debris becomes a rich mulch, cradling many kinds of small creatures, including the eggs and larvae of various kinds of midges. The networks of cavities in the *bastago* pieces retain moisture, even through dry spells, much more effectively than leaf mulch and cacao pod husks.

I timed my annual two or three research visits to Costa Rica for these midge substrate studies to coincide with both the long rainy season and the dry season. At these times I gathered up large samples of the rotting test substrates. My quarry was the larvae and pupae of biting midges. In order to find them, I had to carefully tear open pieces of rotted pseudostems and cacao pod husks. I dumped the contents of artificial bromeliads into little plastic bags, and the ground-cover leaf litter into larger bags. The heavy pseudostems, even after considerable rotting, had to be transported in gunny sacks.

Scooping up bagfuls of leaf litter, pod husks, and pseudostem slices had its occasional surprises. In 1979 I had been collecting large samples of leaf litter from the cacao grove at La Tigra. Flipping over leaves with my machete, I uncovered a nest of eighteen baby fer-de-lances clustered in a moist, bowllike depression. Was the mother around? I gently replaced the leaves covering the snakes and moved on.

Long hours were spent searching through bags of the field-collected organic debris. This was tedious, monotonous work. I used a hand lens to examine immature insects uncovered in the debris, which I picked apart with fine forceps. With the help of Bill Wirth and Ray Gagné of the Systematic Entomology Laboratory at the Smithsonian Institution's National Museum of Natural History in Washington, D.C., leading authorities on the systematics and taxonomy of Ceratopogonidae and Cecidomyiidae, respectively, I was able to learn important field diagnostic characters for recognizing the immature stages of these midges. Over many years I sent samples of larvae, pupae, and adult midges reared from my laboratory cultures, along with wild-caught adult midges, to Wirth and Gagné for identification to genera and species. Specimens were preserved in 70 percent ethanol in small glass vials and labeled by where they were collected. Much of the material eventually was deposited in the permanent collections at the National Museum of Natural History (Smithsonian Institution) and the Milwaukee Public Museum.

In my early work at La Tirimbina, I sorted samples out of doors on a small, rickety table next to the living quarters because I needed maximum sunlight to detect midge larvae and pupae embedded in the dark, soft plant debris. Sorting through the debris also gave me the opportunity to collect a wide variety of other arthropods. This was especially true when examining the pieces of *bastago*. I was amazed to see so many different kinds of tiny creatures, some the size of a pinhead, tunneling through this rotted material. The menagerie within a single piece of rotten *bastago*—beetles, ants, mites, larvae, sowbugs, pseudoscorpions, baby millipedes, tiny spiders, and other creatures I did not recognize (Young 1986a, 1986b)—signaled a vast, little-explored frontier in tropical biology: litter-inhabiting organisms and their biology and natural history.

In sorting through hundreds of these samples, I soon learned that, even though the *bastago* generally contained more immature midges than the other substrates, the majority of pieces of *bastago* did not have midges in them at all. Some pieces contained hundreds of midges and others a few, but most had none. Many hours, over many weeks and months, were spent exposing this patchy distribution of midges in the cacao groves. These data told me something important about the midges, namely, that their breeding populations, while spatially patchy, were spread out thinly over large areas of cacao, and that there were only occasional pockets of high densities of larvae and pupae.

I had set up my initial midge substrate studies with an intuitive notion that I would find a greater diversity of midge species in the La Tigra cacao than in El Uno. I reasoned that the greater ecological heterogeneity of the La Tigra cacao habitat, with its spotty and heterogeneous shade cover of rain forest trees such as *Pourouma*, *Cordia*, *Trema*, *Goethalsia*, and *Cecropia*, would provide a broader range of ecological niches for midge species. El Uno was an older cacao grove, well established with its canopy of rubber trees. It was also more shaded, a condition certainly favorable to midges. Shade cover over the cacao trees in the chosen El Uno sector is about 60 percent at midmorning during the rainy season and about 30 percent during the late dry season. In contrast, shade cover at La Tigra is 20 to 30 percent throughout most of the year.

Flowering in El Uno cacao was somewhat dampened in the rainy season from the heavy shade compared to La Tigra, where flowering was more intense. But the *Hevea* flowers at El Uno were also pollinated by biting midges (Wirth 1956).

Might there be some form of pollinator cross-linkage between cacao and rubber trees in El Uno? While El Uno produced fewer flowers per tree, the ratio of flowers to new pods was similar to that at La Tigra before the testing of midge breeding substrates.

The Sarapiquí substrate study revealed that La Tigra cacao supports a greater number of species of biting midges than El Uno, and that by far the majority of these species breed in rotted slices of banana pseudostems (Young 1982). Midges might be preadapted to breed in the pseudostem pieces based upon the occurrence of rotting pseudostem stalks of wild cousins such as *Heliconia* in light-gap clearings and borders of tropical rain forests. Even species breeding in rotting plant debris in the basal whorled leaf axils of bromeliads might be midges preadapted to exploit banana pseudostem pieces distributed in cacao plantations, already accustomed to occupying small cavities of rotting matter.

Several species of *Forcipomyia*, the genus containing the known pollinators of cacao flowers according to other studies, as well as species of *Dasyhelea*, *Atrichopogon*, and three other genera, were found in the La Tigra pseudostems. And close to seven times the number of midge larvae and pupae were found in La Tigra than in El Uno. While about 70 percent of the immature midges in La Tigra were found in the pseudostems, 20 to 30 percent of the immature midges in El Uno occurred more or less equitably among pseudostems, deep leaf litter, and natural leaf litter. A slight increase in fruit-set was found on tagged trees within a few meters of pseudostem trials in La Tigra, while no such effect was found in El Uno. Very few midge larvae turned up in the artificial bromeliads.

Studies at La Lola

Between 1980 and 1981, as already noted, I designed another midge breeding substrate study for the cacao plantation at La Lola. Repeating the study at La Lola gave me one advantage over the Sarapiquí study. The La Lola plantation contained large plots of fully identified cacao trees, so that it would be possible to repeat my test substrate experiment in areas where the variety of the trees was known. Thus I could deliberately choose areas of the cacao containing self-incompatible varieties such as UF-29, enhancing the likelihood of detecting possible increases in pollination rates and pod-set on trees in the study plots. Such effects are more difficult to ascertain and interpret when self-incompatible varieties are mixed with self-compatible varieties.

Doing cacao studies in another part of Costa Rica also gave me the chance to appreciate the varied topography and rich mosaic of farms and rain forests found throughout this country's mountains and Caribbean lowlands. Turrialba, a small city about a two-hour drive from La Lola, sits nestled within a valley between the mountains south and east of the Meseta Central, about an hour and a half from San José along a twisted and curved mountain road. The fertile mountain slopes, rich with coffee and sugar cane plantations; and the agricultural institute, CATIE, situated between Limón and San José, have given Turrialba an air of agricultural enterprise for many decades.

Accommodations at La Lola include living quarters originally built for banana farm managers and researchers by the United Fruit Company several decades ago. The white wooden two-story structure with red trim is only a short distance from the railway line that hauls bananas and people to and from Limón on the Caribbean coast. Day and night the freight trains come by, those heading east towards Limón brimming with the crescent fruit, still green and ready for the refrigerated holds of freighter ships docked in Limón. To either side of the straightaway portion of track that skirts La Lola there are cacao trees for as far as the eye can see, with drooping reddish boughs forming a crimson layer flush with the raised bed of the railway. Giant canopy trees, left behind when the forest was cleared, are scattered throughout the cacao trees. Toucans and parrots call out with loud squawks from these trees, especially near dusk.

La Lola is an alluring juxtaposition of nature, agriculture, and people. At night, mammoth *Bufo* toads sally forth from the irrigation canals in the cacao to snatch up insects beneath electric lights around the buildings. The scream of the train, piercing through the pounding rains at night, reminds me of Costa Rica's history and the commerce of coffee, bananas, and cacao that forged the economic base of a new nation. Shards of pre-Columbian pottery poke out of the ground in the cacao plantations, bordered with pockets of bananas that attract the flitting *Caligo* butterfly. The jungle makes its attempt to reclaim this land from the cacao and people that have shaped this country's destiny.

La Lola is operated by CATIE in Turrialba, and it is the agricultural institute's chief cacao research station. For many years, CATIE had an active Cacao Program, in which many field studies were being conducted by faculty, visiting scientists, and students at the La Lola facility. Much of the research in the Cacao Program was being funded by the American Cocoa Research Institute. Thus it was logical that my research project would be a welcome addition to studies being conducted at La Lola.

For the substrate study at La Lola, I chose two large separate plots of cacao cultivar UF-29 (United Fruit 29), which is self-compatible for fertilization. One plot was located in heavy shade provided by *Erythrina* trees, while the second plot, about 3 kilometers away, was more open and sunny. This study basically revealed the same trends in midge populations as the Sarapiquí study. The shaded habitat supported a greater diversity of midge species, and the response to the dry season, as evidenced chiefly by the numbers of midge larvae in banana pseudostems pieces, was most pronounced in the sunny habitat, although more species of midges, about thirty total, were found, most of them in the banana pseudostem treatment. Immature stages of midges, especially those of the two most abundant species, *Forcipomyia quatei* and a species new to science, *F. youngi* (named for the author by Bill Wirth "in recognition of his important work on biting midges pollination of cacao"), were similarly abundant in pseudostems during both rainy and dry seasons in the shaded habitat, but much more abundant in the exposed or sunny UF-29 cacao during the dry season (Young 1983c). Subsequently, a similar pattern of concentrated densities of immature midges was detected during the tropical dry season in ground litter in cacao plantations at CATIE, a region of higher elevation than La Lola (Young 1984d).

My discovery of high densities of midge larvae and pupae in the rotted pseudostems during the short dry season at La Lola led me to think about the synchrony of pollinator populations with the seasonally related flowering cycle in cacao. As in the Sarapiquí study, of particular interest were the number of midge species and their distribution by substrate types, and the abundance or density of larvae and pupae of each midge species in the substrates. What eventually emerged from this fieldwork was a picture of the habitat, region, and seasonal patterns of abundance and species numbers for Ceratopogonidae and Cecidomyiidae (Young 1985a). The data generally portrayed a cyclical pattern of midge abundance, in which the greatest diversity and quantity of midges occurred in the wet season. Populations of a few species, especially in the Ceratopogonidae, seemed to be more concentrated (at higher densities) during the dry season in the banana pseudostems.

My second-floor work area and living quarters at La Lola provided a suitable place for rearing the midges from their immature stages in the pseudostems. The large, screened-in room in front of the two bedrooms on the side of the building where I stayed was outfitted with tables I could use for this purpose. I hauled bags of the substrate debris to this second-floor room, filling the floor with many bags waiting to be sorted. I used a small wooden coffee table as my work bench, and a lounging chair as my seat. The little table became cluttered in no time with lots of vials, bottles of ethanol, sorting dish and forceps, and other insect-collecting equipment. The single electric bulb dangling from the ceiling provided only weak light, so I was unable to sort specimens into the night.

Plastic bags in which I reared adult midges from larval and pupal stages in pseudostem fragments were piled onto chairs around the coffee table. Each morning I checked the transparent bags for crawling midges, which I would attempt to trap inside a glass vial. This method worked quite well. Sometimes I left the adult midges in the bags for a couple of days, allowing their cuticle to harden sufficiently to ensure proper condition of the genital structures—parts of the midge that, along with other sclerotized structures, were necessary for Bill Wirth to accurately identify the species.

As with the Sarapiquí study, I saved all of the arthropods I found in the substrates, eventually establishing sizable data sets on the seasonal and habitat distributions of various groups of insects, including ants, staphylinid beetles, earwigs, and others. One of the most interesting discoveries from the La Lola substrate study was the role of rotted banana pseudostem slices as an effective "ecological refugium" for cryptozoic (litter-inhabiting) arthropods, including the immature stages of cacao-pollinating midges, during the tropical dry season. I observed that even during the driest weeks, when the cacao leaf litter crunched loudly underfoot, many pieces of the pseudostem remained remarkably moist on the inside. I interpreted the observed high density of midge immature stages in the pseudostems in the exposed, sunny cacao habitat at this time of the year to be an ecological response to dryness. While immature midges were virtually absent from the leaf litter during this time, their high presence in the rotted, moist pseudostems indicated that the midge population still remained viable. Without the pseudostems—the situation most typical in Central American cacao plantations—there would be the expected severe depression of midge populations as the dry season advanced and a subsequent shortage of pollinators by the time of cacao's peak flowering, which generally occurs shortly after the start of the new rainy season.

A lack of adequate pollination during this time is one of the biggest problems for cacao growers. There is normally a marked asynchrony between the time of peak flower abundance on cacao trees and the peaking of midge populations. My results indicated that adding piles of banana pseudostems to an open, sunny cacao habitat just before the dry season begins could dampen such asynchrony between midges and flowers when the new rainy season commences, two or three months later, by promoting the maintenance of a pollinator population during the dry period. This would be an inexpensive, totally biodegradable method of enriching cacao habitats with midges capable of pollinating the trees.

Most impressive to me about the La Lola findings were the ranges of densities of immature *Forcipomyia* packed into single pseudostem discs. I discovered that the absolute diameter and thickness of the pseudostem slice had little or nothing to do with the number of immature midges it contained. What seemed to matter the most was the degree of moisture in the mulch. The capacity for the pseudostem to retain water in the dry season suggested to me that the slices act as suction pumps, drawing up moisture from the ground cover and retaining it long after the rains have stopped or subsided. In some instances, I found more than a hundred larvae and pupae in one disc, whereas typically in leaf litter samples, I was lucky to uncover one or two larvae of a species such as *F. cinctipes* in 4, 8, or even 16 square meters of substrate, even in the middle of the rainy season, the optimal time for biting midges populations. Even in the obvious presence of ants, predatory mites, and other arthropods, the midge populations in the pseudostems thrived during the dry season, at least in terms of larval and pupal stages.

Pollination Field Studies

But how could I confirm that these particular species were cacao pollinators at La Lola? Clearly further field testing had to be done. It wasn't always clear or confirmed from the literature that particular species of biting midges were indeed effective pollinators of cacao, thus requiring additional study in each instance. I decided to enclose adult midges of the most abundant species obtained from the rearing studies—*F. quatei* and *F. youngi*—into little sleeve cages on flowering branches of UF-29 cacao. From the substrate studies I had identified a trend towards a significant level of increased pod-set on individual tagged cacao trees in close proximity to the piles of pseudostem in the two study plots. By choosing to use a self-compatible clone such as UF-29, I deliberately was stacking the deck to ensure some greater level of pollination on trees within close proximity to the pseudostem piles, compared with levels on trees farther from these areas in the study plots. But statistical significance of differences in numbers of new pods between the two study plots was only at the 10 percent level of confidence, making the observation of increased pod-set on trees closest to the pseudostems limited in scientific value (Young 1983c).

I suspected that adult midges hatching from the piles of rotted pseudostems encountered freshly open cacao flowers close at hand and pollinated them. I had observed that these particular midge species, and several others, during the mid-to-late morning and afternoon hours clung motionless to the staminodes or petals of the flowers. In the rearing studies, I had noticed that adult midges turned up on the sides of the bags chiefly in the nighttime and at dawn, suggesting a hatching rhythm in at least approximate synchrony with the opening of cacao flower anthers and stigma receptivity in cacao—in other words, new adults would be available at a time of the day when cacao flowers were at their freshest. By confining midge pupae from the pseudostem samples along with newly hatched adults into moss-lined cloth sleeves covering the cacao flowers, I was able to confirm that some of the species most prevalent in the pseudostems were, indeed, cacao pollinators. But I was unable to know whether or not I was introducing both sexes into the cages, and in what proportions. This was an important shortcoming due to the possibility that female biting midges are more frequently attracted to cacao flowers and perhaps more significant as pollinators than males. Cacao flowers inside sleeve cages with no midges did not set any pods, while a low but consistent number of pods were set in cages having midges.

In subsequent field studies and assessments of midge pollinator populations at La Tigra in Sarapiquí, *F. youngi* has appeared consistently as the most abundant pollinator species in the cacao plantation, breeding intensely in slices or discs of decaying banana pseudostems experimentally distributed in leaf litter. This research is part of a five-year evaluation study (1989–94) to determine whether or not sustained elevation of pollinator populations in the plantation actually results in a higher level of pod-set and greater annual yields of cacao beans for the small grower.

Adult midges caught on open cacao flowers, as opposed to reared adults, revealed a fairly consistent, intriguing pattern about the relative numbers of males and females. In my field collections of adult midges, and in other published accounts, there is a preponderance of female midges visiting cacao flowers, even

One of the cacao-branch cages used to test if certain species of midges could pollinate cacao flowers in the La Lola cacao plantation in Costa Rica. Midge adults and pupae are introduced into the cage, and it is quickly reclosed. Several such cages are placed in cacao trees. Some receive midges, and others, controls, do not. The flowers inside both kinds of cages are examined daily to determine if any pollination has taken place. Pollinated flowers wilt in a diagnostic manner and do not drop off the branch. In a week or so, it is possible to detect any tiny fruits starting to develop.

though the sex ratio at birth (emergence from the pupa) is about one to one. This observation suggests that female midges seek a specific kind of nourishment, perhaps certain amino acids from the high-protein pollen grains, directly linked to their reproductive physiology, in much the way that female mosquitoes feed on blood for egg-building amino acids.

Midges also tend to be most active at dawn and dusk in plantations, perhaps coinciding with peak floral receptivity for pollination. To determine if this was so, I measured the temporal pattern of flower opening by "marking" flowers (I stuck color-capped pins next to floral cushions bearing large buds) so that I could track their individual patterns of opening at regular intervals throughout the day and night (Young et al. 1984). I was able to confirm what a Dutch scientist (Wellensiek 1932) determined many years before: The cacao flower begins to open gradually in the late afternoon, and continues through the night, so that the flower is fully open just before dawn the next day. This daily cycle of flower opening and maturation (Young et al. 1987a) was important in understanding the association of pollinators to the flowers.

In 1983, Melanie Strand, a student working on her master's degree at the University of Wisconsin in Milwaukee, helped to clarify the temporal pattern of receptivity of the cacao flower to pollination. Working at CATIE in Turrialba, she collected fresh cacao flowers at different times of the day and night, including mature buds about to open, and carefully applied hydrogen peroxide to the stigmatic surfaces. Bubbling indicated stigma receptivity to pollen grains, that is, initiation of the complex pollination process. We discovered that the stigma of the cacao flowers, as well as those of other *Theobroma* and *Herrania*, were very receptive for pollination during the early morning hours (Strand 1984; Young et al. 1984). We also found that the pollen sacs on the anthers split open in the morning to release pollen grains, making them available for insects to pick up. Since cacao pollen remains viable for at least twenty-four hours, a pattern of dawn-dusk visitation to the flowers by insects would provide an ample window of opportunity for pollen retrieval (dusk to dawn) and deposition (dawn and morning hours), bringing about effective natural pollination (Young 1985d; Young et al. 1987a).

This pattern of stigmatic receptivity and pollen availability is known as protogyny—the flowers are functionally female first, and then male (Young et al. 1987a). Protogyny may have evolved to prevent self-pollination, which would interfere with the deposition of potentially more adaptive pollen from other individuals. Likewise, self-pollination may compete with incoming pollen for optimal germination sites in the pistil and ovary. Furthermore, outcrossing frequently results in higher seed quality and reproductive success. By displacing the female

FLOWER CUSHION

FIRST STAGE: Insect feeding on staminodes

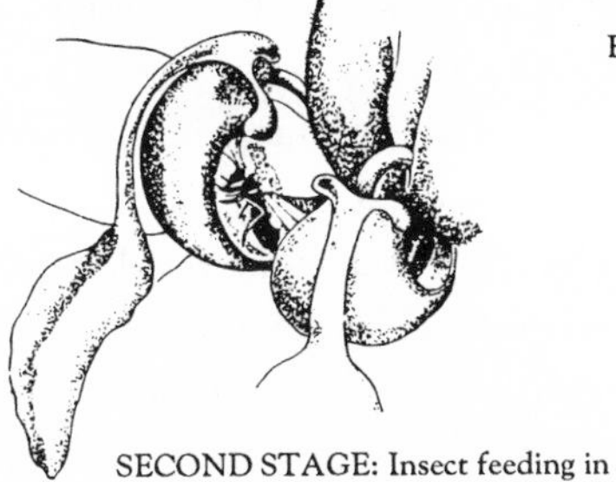

SECOND STAGE: Insect feeding in petal cup and gathering pollen

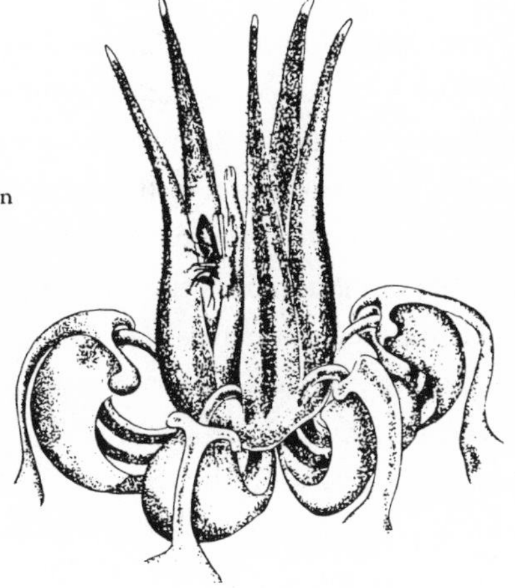

THIRD STAGE: Insect feeding in other flower on staminodes and spreading pollen on pistil

Pollination of the cacao flower by the ceratopogonid midge. From Bystrak and Wirth (1978).

and male functions of the flower temporally, such interference is minimized. This is especially adaptive in species bearing closely aggregated flowers, such as cacao, since the probability of potentially deleterious stigmatic clogging increases and the mechanism of pollination is less precise. Protogyny, therefore, is particularly adaptive in self-incompatible species and thus contributes to the success of cross-pollination.

At about the time these floral studies were initiated, I visited the laboratory of Kamaljit Bawa at the Boston campus of the University of Massachusetts to learn the laboratory technique of fixing and staining pistils from cacao flowers to check for pollen tubes in various stages of growth. A pollen tube grows through the pistil tissue towards the ovary when pollination has been successful and allows for the fertilization of the egg cells (ovules) in the ovary by the pollen. In most plants, only the stigmatic surface at the top of the style is involved in pollination. In cacao both the stigma and style are sites for pollen transfer and germination of the pollen tubes.

Cacao is also unusual in that the site of compatibility is in the ovary rather than in the pistil, as is typical in most plants. Pollen tubes, one for each pollen grain germinating on the stigmatic surface of a flower, may grow all the way down into the ovary before the compatibility system in cacao expresses itself and allows fertilization or not. I wanted to take advantage of this condition to measure or assess the daily cycle of pollen-tube growth in cacao as an indirect measure of optimal pollination times. As Bawa explained to me, one can extrapolate the time of pollination from the stage of pollen tube growth.

I had collected samples of pistils from cacao flowers at La Lola in various stages of opening and throughout their attachment to the trees. These samples were stored in formaldehyde-acetic acid solution, typically used for preserving plant tissues. By preparing the samples in the laboratory using specific chemical reagents, I was able to examine the pistils, squashed on microscope slides, with a fluorescence microscope. By so examining large samples of flowers, we were able to strengthen our conviction that cacao flowers are most likely pollinated at dawn and dusk, as my field observations and analyses of anther dehiscence had indicated.

I wanted further data on the activity patterns of midges in relation to the cacao flower activity cycles, as well as more data on the adults, to corroborate the information I was amassing on the immature stages. Yet the adult midges, when flying through the cacao, were not much bigger than tiny specks of airborne dust and extremely difficult to detect. In 1982 I devised a means of trapping them in close proximity to flowers, using small rectangular pieces of sticky flypaper, each about 4 by 2½ centimeters. I pinned these pieces of *papel matamoscas* around each of many freshly open cacao flowers on several trees at La Tigra, forming a sticky boundary within millimeters of each flower. I hoped to snare small flying insects as they zeroed in on the flowers. By leaving the flypaper out for several twenty-four-hour cycles, and checking them throughout the day and night, I would be able to detect a possible diurnal cycle in insect abundance at the flowers. I used forceps to extricate the insects, or pieces of insects, such as snared legs and wings, from the flypaper and placed these in vials of alcohol for examination at a later time.

On one tree, I stripped off all the flowers and floral buds but pinned pieces of flypaper to the branches to determine if flying insects were as abundant on a cacao tree when flowers were absent. After several nights, I could detect no difference in trapped insects between the flowering trees and the deflowered tree, perhaps an artifact of small sample size or the close proximity of the stripped tree to the others bearing flowers (Young 1986c). Later, in the same month (July), a time of high flowering in cacao, I repeated the exercise at La Lola. This yielded an impressive seventeen families of Diptera, including a small contingent of biting midges at La Lola and La Tigra (Young 1986c, 1986d). What was especially surprising to me was that about half the total samples for both localities combined were phorid flies—the pollinators of cacao's allied genus, *Herrania* (Young 1984e).

At the time of this discovery, I was already taken with the unusual flowers of *Herrania* trees in a garden plot at La Lola. Although similar in floral design to *Theobroma*, *Herrania* flowers possess greatly elongated petal ligules, in some species appearing as slender ribbons drooping downward. Some species bear purplish, maroon, or dark red flowers, while at least one bears white flowers. One of the things that all species of *Herrania* have in common is a musky, rotten meat smell, clearly indicative of fly pollination.

Not surprisingly, I found that *Herrania* flowers are intensely visited by the stout-bodied phorid flies at dawn and again near nightfall, and they presumably pollinate the flowers (Young 1984e). The phorids, owing to their bigger body size, were much easier to observe than the tiny biting midges or gall midges associated with cacao and other *Theobroma* species. The phorids, all females, showed up at the flowers like clockwork each dawn, entering them swiftly and exiting with a thick down of yellowish pollen on their bodies. Sometimes the flies alighted on the long petal ligules and crawled up them to reach the concealed pollen sacs. Flies rummaged around intently in the entire flower, brushing against the pistil and stigma. Female phorids might be fooled, attracted by the odor of *Herrania* flowers, similar to that of their usual egg-laying sites in nature, namely, decaying animal and plant matter.

The high abundance of these flies, which I now assumed to be the chief pollinating insects of *Herranias* at La Lola, was consistent with my observations that these trees bore large numbers of pods, as if pollination success was close to 100 percent. Examining collected specimens of the flies, I noticed their bodies to be richly covered with bristles, ideal for snaring sticky pollen grains. The pollen grains of *Theobromas* and *Herranias* are coated with a sticky adhesive and are not well adapted to passive dispersal by wind or rain droplets.

These side studies provided a much needed therapeutic interlude from the monotony of sorting through midge breeding substrates in the cacao pollinator studies. They also helped me to understand cacao in relation to its closest evolutionary relatives, perhaps providing further insight into the mystery of cacao pollination itself.

Gall Midges

Ceratopogonids were not the only kind of midges I had been seeing on cacao flowers at La Lola and in Sarapiquí. Gall midges of the dipteran family Cecidomyiidae were also abundant on cacao flowers. I had read a report that one species of gall midge, of the genus *Aphodiplosis*, was possibly a pollinator of cacao in Ghana (Kaufmann 1973). Other reports, chiefly those from Brazil, generally dismissed gall midges as cacao pollinators in spite of their much greater abundance on cacao flowers than biting midges. In La Lola I encountered many gall midges clinging to cacao flowers during the daytime. They were most abundant in the heavily shaded cacao. I started collecting samples of the adults by trapping them in glass vials.

I noticed that these delicate, tiny insects would collectively "roost" on spider webs in the cacao. Although it was known that gall midges commonly rest on spider webs in the tropics, it was particularly amazing for me to witness this intriguing behavior, since at least some kinds of spiders feed on these midges. I had seen jumping spiders, which do not build webs, skillfully capture gall midges alighting on a branch or leaf in a cacao tree.

Gently blowing on a web laden with the midges, I found that I could not easily dislodge them. Tapping on the same web with a twig or forceps sometimes caused the group to disperse into the air and resettle moments later on nearby leaves and branches. Returning to the web an hour or so later, I found a cluster of midges, possibly the same group as before, pretty much reassembled on the same web. Intrigued, I searched the cacao trees for more webs, marking their locations and documenting the presence of gall midges. From the samples I had been collecting from cacao flowers and from the webs, Ray Gagné was able to identify what the various species were. To collect a sample from a spider web, I slowly enclosed the web in a widemouthed jar, then capped it quickly. I was surprised to learn that many of these samples contained mixed species of gall midges rather than one.

Not all webs in a cacao tree or in adjacent trees are occupied by midges. Why is it that certain webs, and not others, are selected? Since I learned that midges roosted in webs constructed by different kinds of spiders, including members of the Theridiidae and Araneidae that turned up in my field samples, it was unlikely that web preference by midges was based upon the kind of spider. I also observed that midges left the roost sites exactly at the same time of day that midges resting on the open cacao flowers also took flight, namely, around dusk. Midges reappeared on the spider webs and cacao flowers shortly after dawn. They did not return to the webs during the night. Clearly these gall midges exhibited a rigid diurnal activity pattern in the cacao.

During the dawn hours, and just after sunrise, I observed that some gall midges, which turned out to be chiefly what I called *Mycodiplosis* "sp. 1" and *Clinodiplosis* "sp. 1," moved through freshly opened cacao flowers, rather than just perching motionless on the petals or staminodes as they often did later in the morning and all afternoon. I discovered that these midges moved through the flower in a manner very similar to that of biting midges, suggesting that gall midges might also play a role in cacao pollination.

I therefore collected some specimens from the flowers, heat killed them by putting the vials in direct sunlight, and, with the help of Melanie Strand, examined them with the electron microscope at the University of Wisconsin-Milwaukee. This survey revealed that gall midges are not as well endowed with dense, sturdy bristles as are many biting midges. I was able to find only one pollen grain adhering to some long filamentous abdominal hairs on one midge. Perhaps in the wild, gall midges are not well equipped to pick up and transport cacao pollen grains on their smooth, slender bodies.

Observing gall midges in the cacao brought into focus more questions than answers. I discovered that *Mycodiplosis* sp. 1 was one of the most common visitors to cacao flowers. During March 1982, I counted twenty-five midge roosts in fifty cacao trees in the heavily shaded UF-29 plot, representing a total of 723 gall midges, mostly *Mycodiplosis* sp. 1. Eventually this midge was described as a new species, *Mycodiplosis ligulata* Gagné, its species designation derived from its unusual mouthparts (Gagné 1984). In most gall midges, fluids are usually lapped up from an open feeding substrate, but this is not the case with M. *ligulata*. Owing to its slightly elongate proboscis, this species can reach nectar or other nutritive fluids through long, narrow openings. It is possible that cacao flowers secrete minute quantities of an adhesivelike nectar coating the lower reaches of the staminodes and ovary area (Strand 1984). Midges would have to access these fluids through narrow passages on the flower. Could this particular species of gall midge have a specialized association with cacao flowers? I had also found gall midges, including species of *Mycodiplosis*, visiting the flowers of *Theobroma simiarum* and *T. mammosum* at La Lola (Young 1985c).

Once, I witnessed a gall midge perform a very curious behavior in a cacao flower (Young 1985c). This event occurred in the afternoon, about half an hour before sunset. I had been standing about 10 inches in front of a flower for a half an hour watching what I later found out was a female *M. ligulata*. The gall midge was clinging to the petal ligule and occasionally moved its head downward, its mouthparts coming into contact with the soft, yellow petal tissue. My feet were tired, and, in the heavy shade, it seemed as if dusk had already arrived. I avoided swiping at the no-see-ums buzzing around my eyes. Suddenly I saw the gall midge flip itself around and enter the little cavity of the petal pouch. The petal hood vibrated as the midge rumbled around inside, momentarily changing the shape of the pouch. The midge came out quickly and flew away. This was the first time I had witnessed a sequence of gall midge movements and behavior inside a cacao flower. These movements of the gall midge in the open cacao flower suggested to me a specialized and orderly interaction between this insect and cacao, highly suggestive of a specialized pollination mechanism.

Although I never witnessed this behavior again, I did observe gall midges alighting on the staminodes of cacao flowers and crawling up towards the ovary, compressing themselves in the narrow space between the pistil and staminode as they moved along. I had also noticed that the tips of the staminodes on freshly opened cacao flowers are remarkably flexed or curved inward, all five pointing towards the central stigma beneath them. Could it possibly be that once a midge crawls into the narrow space between the staminodes and pistil it is momentarily trapped in a "prison," making it difficult for the insect to turn around or otherwise back itself out of the flower's reproductive core? I was compelled to think about the unusual biting midges pollination system in *Aristolochia* vines, in which the midges are trapped for a couple of days in a bulb-like floral pouch until pollination takes place. Specially designed and strategically placed hairs prevent the midges from escaping, and only when these "prison bar" hairs wither can the insects exit the flower. *Aristolochias*, like cacao, produce midge-attracting musty floral aromas. Pollinated cacao flowers exhibit recurved or outwardly flexed staminodes, allowing easy escape of midges. Other midges land on a staminode or petal ligule and then slowly crawl into the petal hoods. Thus in two very different plant groups (*Aristolochia* and *Theobroma*), there are some striking similarities in floral design and behavior to accommodate small-bodied insects as pollinators. In this context, it is interesting to note that the flowers of *Aristolochia pilosa* in Panama are pollinated by females of certain small-bodied dipterans (Wolda and Sabrosky 1986), although other *Aristolochias* are pollinated by females of larger phorid flies of the genus *Megaselia* (Hall and Brown 1993), the same genus pollinating *Herrania* in Costa Rica (Young 1984e).

What was impressive, too, as I had discovered with biting midges, was the marked preponderance of female gall midges on cacao flowers. This was especially true of the two most frequently encountered gall midge species, *Mycodiplosis ligulata* and *Clinodiplosis* sp. 1 (Young 1985c). From information available on other gall midges, it was safe to assume that the sex ratio typical for populations of most species is one to one. Were these gall midges obtaining something from cacao flowers central to their reproductive biology? Only continued research on neotropical gall midges biology will be able to eventually provide the answer. While still inconclusive, my observations suggest that female gall midges in particular visit cacao flowers to feed on pollen as a source of egg proteins, and in so doing, pollinate the flowers.

These observations prompted me to confine wild-caught gall midges in the sleeve cages described earlier to determine if they could pollinate cacao flowers. I followed the same procedure as for biting midges, but I could not control the sex of midges introduced into the cages. I was fairly certain that I could recognize specimens of M. *ligulata* in the field and use it in the cage tests. Undoubtedly, I made some errors in identification, but the exercise revealed that pod-set did take place in cages where gall midges had been introduced. While I did not actually witness the acts of pollination, given my observations of these midges on unconfined cacao flowers, I am reasonably sure that these insects were the pollinating agents in the cages.

Finding the larvae of gall midges to understand their breeding substrate requirements proved far more difficult than locating those of biting midges. Midges such as *Mycodiplosis* feed on fungi as larvae and do not form galls, but it was not a simple matter of finding galls on leaves or twigs and discovering larvae inside. I had many places to look, including the ground cover of cacao and the meshwork of epiphytes and debris draped from branches of old cacao trees in the heavy shade. Then I started noticing mature cacao pods with large gaping holes chewed by squirrels. On some of these, there was a fine carpeting of fungal growth, a grayish mat of fluffy hyphae coating the exposed cut walls of the hanging pod and issuing from the soggy contents inside. I could also see gall midges alighting on the hyphae, perhaps to lay eggs. Biting midges were not present.

Breaking open these pods, I found the pink larvae of *Mycodiplosis* and other gall midges on the fungus-coated surface of the pod. On one occasion, when I was able to take a sample of the fungus from a wounded pod to the phytopathology laboratory at CATIE, I found out that the "fungus" was actually a mixture of three different kinds: *Fusarium*, *Phytophthora*, and *Theilaviopsis* (Young 1985c). These fungi are early colonists in the fermentation and decay of the pod's inner tissues, once exposed by pulp-seeking squirrels.

Gall midge pollination is relatively unknown, or at least poorly documented, anywhere in the world. The data are still inconclusive on the overall effectiveness of gall midges as pollinators of cacao. However, there is enough information now available to indicate that further research in this area is needed. The best documented case of gall midge pollination is a recent study in Ecuador involving small trees of the genus *Siparuna* (Monimiaceae) (Feil 1992). Thirteen species were studied, and their dioecious flowers are specialized for pollination by gall midges. Male flowers in these species are especially designed morphologically to encourage egg-laying by gall midges, during which loads of sticky pollen are picked up on their bodies when they insert their abdomens through a pore in the flowers. Midges exhibit similar behavior at female flowers, which are less accessible as oviposition sites, and pollen is brushed off their bodies onto stigmas, causing pollination to occur. Midges are attracted to these nectarless flowers by a lemon scent, and their larvae complete their development chiefly in male flowers. The flowers are outfitted with a landing platform area to encourage gall midges to alight on them. Various species of *Theobroma* emit lemon-like floral scents, and the flowers, while very different from those of *Siparuna*, have landing platforms for gall midges. While structurally different from *Siparuna* flowers, *Theobroma* flowers are designed so that accessibility to both pollen and stigmas, as in *Siparuna*, is difficult for midges and other insects. Whether or not gall midges lay eggs in *Theobroma* flowers, including cacao, has not yet been studied.

Midges in Belize

During these field studies, I had the opportunity to observe animal-damaged mature cacao pods at another site in Central America, the Hummingbird Hershey Cocoa Farm in Belize. I had first visited Hummingbird Hershey during August 1981, the rainy season, at the invitation of B. K. Matlick and Gordon Patterson, both of Hershey's Agribusiness Division. My mission on that first visit was to survey the cacao farm for biting midges, including their breeding substrates, my chief interest at the time. For both visits, I flew from Costa Rica to Belize on the early morning "milk run" SAHSA (Servicio Aéreo de Honduras, Sociedad Anónimo) flight, which stopped in Managua, Nicaragua, and Tegucigalpa and San Pedro Sula in Honduras before touching down at the little airport on the outskirts of Belize City. These visits to Hummingbird Hershey were a means of expanding my experience-base with cacao in Central America.

The main purpose of establishing the Hummingbird cacao farm was Hershey's desire to provide a model cacao plantation in Belize. Hershey hoped that by encouraging Belizeans to grow cacao, the crop would provide a new successful economic base for the country. It was also expected that the farm would provide Hershey with a sizable supply of cacao beans. Cacao had been farmed at the site long before Hershey acquired it in the late 1970s. The original forest had been cleared much earlier, planted in cacao by some Jamaicans in the 1950s, but abandoned by 1962. The farm was eventually sold in 1970 to a Canadian real estate developer, who in turn sold it to Hershey in 1976.

On my second visit, during the dry season in March 1983, I also set out to survey the breeding substrates of gall midges. I noticed that diseased pods, rotten and blackened but still hanging on the trees, were a prime breeding site for *Forcipomyia* midges, here as in Costa Rica. I scraped away the tough pod tissues of the cherelles and larger cacao pods and found little pockets of the pink larvae of various species. I noticed many woodpecker-damaged mature pods and found gall midge larvae living inside. Hummingbird Hershey thus helped broaden my perception of the breeding sites of these potential pollinators. I started to develop a composite view that damaged and diseased cacao pods are suitable microhabitats for the breeding of various kinds of midges of possible benefit to cacao production—a view somewhat in contrast with the generally held perception that diseased and damaged pods should be cleaned out of the plantation as a sanitary measure against further infections of cacao diseases.

Habitat Studies

Upon returning to Costa Rica, the main site of my cacao research, I began to solidify my perceptions of how cacao pollinators benefited from the availability of rotting cacao pods as breeding sites in plantations. In wild populations of cacao, I envisioned an evolutionary scenario in which animal-damaged mature pods provide an important arboreal breeding niche for pollinating midges, which, upon completing their life cycles, would be available in close enough proximity to the flowers to effect pollination. My thoughts were reinforced by observations of the Matina cacao forest in an abandoned grove at La Lola. The grove appeared, for all practical purposes, as if it had reverted to rain forest. Every time I drove or walked past this cacao forest, I grew more intrigued by it.

Across the road from the cacao forest stood a well-maintained parcel of Matina cacao. These too were old trees, but short and squat, nursed by decades upon decades of pruning. The abandoned cacao trees were tall and slender, bearing high branches free of epiphytes under the shaded canopy. But this was not the case in the maintained La Lola Matina cacao stand. Here, the twisted, thick branches of the cacao trees were festooned with mistletoes, tiny pendant vines, ferns, and much more. Seeing the great contrasts between these two stands, about 800 meters apart, of Matina cacao ignited my interest in surveying their biting midge and gall midge populations. I was curious, too, about the high incidence of squirrel damage to the pods in both areas. Here was a chance to perhaps learn more about the arboreal breeding sites of cacao-associated gall midges and biting midges.

Many of the pods in the cacao forest were stained with the handiwork of *Monilia* and *Phytophthora* diseases. Abandoned cacao plantations in the Atlantic zone are reservoirs for fungal organisms that attack cacao pods. In the pristine rain forest, the fungi would just be a small part of a complex balancing act imposed by networks of specialized predator-prey associations, which dampen the tendency for most species to proliferate. But vast monocultures of cacao are sitting ducks for sweeping infestations of such organisms. From casual observation, I had noticed that there appeared to be more squirrel-damaged pods in the maintained Matina cacao than the Matina cacao forest, but that the reverse was true of diseased pods. Squirrels did not seem to attack mature pods bearing obvious signs of primary fungus damage. Of course, squirrels, in chewing into pods, created conditions suitable for secondary damage to pods by fungi.

Given the often severe damage to cacao pods from squirrels, woodpeckers, and fungi, it would seem that there is little adaptive value to pods remaining on the tree and being killed. But this apparent disadvantage must be offset by a strong advantage in having healthy seeds dispersed and scattered in the wild by monkeys and other animals. Given the conservative breeding strategy of cacao, perhaps only a low number of seeds need to survive and be dispersed to maintain wild populations. Again, vegetative proliferation, with the production of multiple stems or trunks, may be more important than regeneration of wild populations from seed.

From October 1983 to December 1984, I began studying the breeding sites of midges in these two distinctive stands of Matina cacao in relation to squirrel damage and fungal disease of pods. Twenty more or less randomly selected trees in each area were tagged with yellow and red plastic strips and checked each month for the numbers of flowers and pods. These measurements were taken during the entire fourteen-month study. I arranged with local workers to distribute banana pseudostems, cacao pod husks, and leaf litter around the sites, using the basic methodology developed for the previous studies. This experimental design would allow me to detect any midge breeding preferences for certain types of rotting debris over others and learn whether or not cacao trees nearest to higher numbers of midges (as determined by recording their larvae and pupae in the debris treatments) tended to have higher levels of pod-set, reflecting higher pollination levels (Young 1986e).

The number of diseased pods and small pods killed by cherelle wilt was also recorded. In addition, I counted the abundance of immature midges in thirty-five full-sized rotten pods on trees in each area and another thirty-five pods rotting on the ground, which had been cut off the trees because of signs of *Monilia* and *Phytophthora* disease. *Monilia*-rotted pods bear large, roundish blotches of powdery white or yellowish mold on the surface, telling of decay in process on the inside. *Phytophthora*-rotted pods have large, blackish blotches. On my visits to La Lola during the study period, I also counted the number of squirrel-damaged pods on the tagged trees in both areas. My findings confirmed the inverse relationship between squirrel-damaged pods and fungal-diseased pods in Matina cacao (Young 1986e). Squirrels killed many more pods in the well-maintained, open, sunny stand of Matina cacao than in the cacao forest, even though pod numbers between the two areas were similar for the entire study. Also, many more pods were killed by fungus in the forest. Gall midges represented by four genera, including *Clinodiplosis*, were found breeding in the disease-killed pods on the ground and in the trees, whereas twelve species of biting midges were breeding in the rotten pseudostems and, to a lesser degree, in cacao pod husk pieces in both areas.

I also discovered that cacao trees situated beneath shade trees in the open La Lola habitat tended to have greater pod-set than cacao trees away from shade trees in the same area. But pod-set was much more uniformly dispersed across trees in the heavily shaded cacao forest. These observations tell me that pollinating midges preferentially are more active in shaded areas of open, sun-drenched cacao than in more exposed areas. This pattern is somewhat analogous to a situation in Peru. Bob Hunter observed that cacao there is planted in steep stream gullies in the hills near the Pacific coast, a very dry region. The trees closest to the stream, where it is damp, have many more pods than trees further up the slope, where it is dry and open. Pollinating midges are most likely limited in their movement patterns by the availability of shade and moisture.

Where the cacao meets the rain forest, trees have many more pods than those more centrally located in the plantation. I have observed this striking pattern over several years at La Tigra. The well-shaded cacao trees closest to the beginnings of the rain forest, which bear many pods in various stages of growth, are festooned with vines and epiphytes. Underneath these trees, the leaf litter is deep and spongy—ideal breeding conditions for cacao-pollinating midges. Some species of midges breed in the moist debris lodged in bromeliads, orchids, and the other epiphytes densely blanketing the cacao trees bordering the rain forest (Young 1986d, 1986e). In the central part of the *finca*, where the shade is patchier and less dramatic, there are fewer epiphytes and vines about the trees.

The rain forest itself, with its plethora of rotting debris, dense shade, and variety of plants, is the reservoir of cacao-pollinating midges (Young 1986d), which most likely exploit many different kinds of food sources. *Theobroma* flowers might be much more directly dependent on these midges or gnats for pollination than vice versa. Specialized flowers often attract only certain kinds of animal pollinators, but the animal pollinators interacting with these specialized floral types are most likely ecological generalists to some degree—that is, their survival and reproduction is not solely dependent upon these particular floral resources. I believe that *Theobroma* species, including *T. cacao*, fall into this category, having intricate and specialized flowers that attract classes of insect pollinators most likely having unspecialized feeding habits. I would add one qualifier, however. Given the skewed sex ratio of midges trapped in cacao flowers, leaning heavily towards females, cacao-pollinating midges probably obtain specific nutrients from cacao flowers for the development of their eggs, or some other substance important to the breeding process, although this hypothesis remains to be studied.

Fragrance Studies

Although I was finding high densities of immature midges in rotting banana pseudostems at La Lola and in Sarapiquí, all of my observations at La Tigra and El Uno suggested generally low numbers of adult midges in cacao habitats. This was especially true of ceratopogonids, rather than gall midges. For both groups of midges, I found a low frequency of adults in the act of pollinating cacao flowers, as well as other *Theobroma* flowers. This dilemma, which has plagued cacao pollination studies for almost a century, prompted me to develop another avenue of research. I elected to study the external morphology and related floral attraction-reward system of *Theobroma* and *Herrania*, including cacao. To get started, I sought out expertise I did not have. With the cooperation of Marilyn Schaller and Melanie Strand at the University of Wisconsin in Milwaukee, I set up a scanning electron microscope (SEM) study of *Theobroma* and *Herrania* flowers.

I collected flowers of *Theobroma* and *Herrania* species in Sarapiquí and La Lola and brought these back to Milwaukee in the appropriate fixatives needed for SEM studies. In the three species we examined—*T. cacao*, *T. speciosum*, and *H. purpurea*—we were unable to locate discernible quantities of nectar in fresh flowers. One of the first things we noticed was the marked absence of any coatings that would have been floral nectar. This corroborated our field experience of being unable to detect or collect any appreciable amounts of nectar from the living flowers. Based on our observations, we tentatively concluded that nectar, if present at all, exists as a tightly bound, adhesivelike film on the surface of specific floral structures, rather than as a free-flowing copious fluid collecting in the flowers.

The SEM studies revealed glandular, clublike structures coating the ovary surface in all three species (Young et al. 1984). Most interesting, we observed a conspicuous "basal ring" of densely packed multicellular trichome hairs between the sepals and petals at the base of the flower. This structure is common in the Malvales, the order to which the Sterculiaceae belong. While this ring-like structure is continuous in some species such as *T. speciosum* and *T. mammosum*, and in the related genus, *Herrania*, it is markedly broken and fragmented in cacao. Could it possibly be that the strong but differing fragrances of *T. speciosum*, *T. mammosum*, and *Herrania* are emitted from these glandlike trichomes of the intact basal ring structure? And could it be that the barely perceivable fragrance typical of cacao is a result of the fragmented and perhaps dysfunctional ring we observed in this species? Our findings compelled us to delve deeper into the floral biology of *Theobroma*.

Our work soon evolved into an effort to determine the nature of floral fragrance in cacao and related species. Strand accompanied me to Costa Rica in 1985 to collect additional floral material for further SEM and TEM (transmission electron microscopy) studies, and for steam distillation of the floral fragrance "oils"—our first attempt to discover the biologically active substances in *Theobroma* and *Herrania* floral fragrances that possibly modulated pollinator behavior. Using TEM, Strand discovered the presence of large amounts of lipids in the basal ring glands. Using a lactophenol stain she demonstrated an elaiophor-like function in these structures—that is, they produce volatile oils. Strand's thesis research and our subsequent studies highlighted the great variety of floral size, shape, color, clustering, and fragrance exhibited by different members of *Theobroma*.

Strand had been in touch with biologists such as Norris H. Williams of the Florida State Museum, a leading authority on the biology of neotropical orchids and their pollinating bees. Williams had developed techniques for studying floral fragrances. We also journeyed to the nearby University of Wisconsin campus in Madison to confer with Barbara Erickson and Eric H. Erickson, Jr. Eric and his students were studying the floral attraction-reward system of soybeans and sunflowers and other plants in relation to honey bees as pollinators, and Barbara had completed a doctoral thesis exploring the nature of floral fragrances and other volatiles in cotton plants.

After receiving helpful advice and training from Barbara Erickson, Strand put together a plan for a vacuum steam distillation of *Theobroma* and *Herrania* species' floral oils. We set up a makeshift distillation system at the laboratory facility of the Cacao Program at CATIE in Turrialba. The Ericksons loaned us a portable distillation set of glassware packed in two metal suitcases, which we brought to Costa Rica along with several boxes of other supplies. I located, with the help of Ludwig Mueller at CATIE, an unused "Roto-Vap" condenser, crucial for the distillation process. José Galindo lent us his new water bath apparatus. We brought large steel cylinders of nitrogen gas from San José, tying the big, heavy containers onto the back of a jeep.

Borrowing from what other researchers had done with orchid fragrance studies, we also attempted to collect samples of floral fragrances using an absorption-on-Tenax-cartridges technique. In this approach, air is drawn by suction over enclosed samples of fresh flowers, and the fragrance volatiles are presumably collected on the absorbing surfaces of the cartridges. In our studies, this approach proved most futile, perhaps because, as we eventually discovered, there was relatively little fragrance with which to work in *Theobroma* and *Herrania*.

Therefore, we decided to set up a distillery to boil freshly collected flowers from the cacao groves as well as from *Theobroma* and *Herrania* gardens on the CATIE grounds. In this approach, a known quantity of freshly collected flowers is subjected to one hour of distillation at 100 degrees C. The distillate is then extracted three times with a solvent (chloroform or dichloromethane) and dried over sodium sulfate. The tiny amount of concentrated floral oil obtained was then frozen and stored for gas chromatography and mass spectrometry analysis at the Wisconsin State Laboratory of Hygiene at the University of Wisconsin campus in Madison, with the cooperation of Doug Dube.

This is not to say that steam-distilling floral oils was the best way to get a picture of the substances that made up the floral fragrances. It would have been much better had we been able to collect fragrances as they were emitted from flowers in situ and to analyze these samples for chemical composition. With distillates, there is a strong likelihood of picking up breakdown products and other substances along with the fragrance. So we had to appreciate the limitations of our approach and be careful in interpreting the results of the gas chromatography and mass spectrometry that followed our distillations in Costa Rica.

In our initial study, we collected cacao flowers from the cultivar UF-613 at different times of the day and night to determine if there was a definitive change in floral fragrance chemistry over the day-night cycle. In this study and a subsequent one, we included an examination of several species of *Theobroma* in addition to *T. cacao*.

What we could not collect for distillations from the CATIE garden, I obtained in the *Theobroma* garden at La Lola. These samples of flowers were carried in Zip-Loc bags or sleeves of aluminum foil. In between distillations, we experimented with ultraviolet and infrared photography of cacao flowers and conducted a bagging study in which we determined, to no one's real surprise, that cacao flowers deprived of contact with pollinators fall off the tree within a day or two after opening. Gradually we were able to obtain enough clean samples of the floral oils to do an adequate analysis of major volatile constituents. I transported the floral oil samples back to Wisconsin using a small ice chest or cooler I hand-carried on the plane.

Once ion chromatograms had been obtained from our 1984 and 1985 distillations, Barbara Erickson spent long hours identifying the major volatile constituents in our samples—the components of the floral oils that would most likely be involved in the attraction of insects to the flowers. From what is known about floral fragrance in other plants, a combination of these substances usually provides an attraction cue for pollinators, not just one or a few principal components of the floral oil. Gas chromatography and mass spectrometry permit not only the determination of constituents, but also their relative and absolute abundance in the samples.

Contaminants invariably turn up in samples, complicating the analysis. Even though our floral oils were stored in special sealed vials, there was no guarantee of completely preventing contamination during the several thousand miles of transport. And there was always the possibility that the samples were contaminated during the distillations, given the makeshift nature of our setup and the less-than-ideal conditions for this sort of research, although over the years our distilling technique improved, and we produced better, contaminant-free samples (Young et al. 1987c, 1988; Young 1989a, 1989b).

Still, our results gave us some interesting insights into the nature of floral fragrance in *Theobroma* and *Herrania*, and especially how the composition of floral oils varied among different species (Erickson et al. 1987). We discovered that the floral fragrance oil of cacao consisted chiefly of saturated and unsaturated hydrocarbons. Of the seventy-eight substances identified in the oil, pentadecene and pentadecane were the principal constituents, the latter comprising about 50 percent of the total by volume. We also found that pentadecene was present in the floral oil at similar concentrations at night and at early and late mornings, but had dropped to a low of about 1 percent by late afternoon. This observation suggested to us that, under the assumption that pentadecene, the major volatile constituent of cacao floral oil, is an attractant for pollinators, it is most abundant in fragrance during the morning and again in the early evening, but not in the afternoon—a pattern somewhat consistent with the activity of pollinating midges chiefly in the morning and at dusk, and possibly after dark. At the same time, however, we also appreciated the fact that volatiles present in much smaller quantities could also be contributing to the overall attractant properties of this floral fragrance for pollinators.

Perhaps the most startling find was that cacao flowers emitted such a chemically complex fragrance in the first place. Certainly cacao flower fragrance was not very noticeable in the field. Yet we now had clear evidence of a whole suite of volatile oils underlying the fragrance of cacao flowers. This had never been reported before. Our discovery of what Barbara Erickson would term "oily hydrocarbons" as the main constituents of cacao floral oil was consistent with our perception that these flowers have a musty, almost acrid aroma after being enclosed in a vial (Erickson et al. 1987).

Of the fifty-eight compounds we identified from the distilled floral oil of *T. mammosum*, the linalool oxides, which were not found in cacao, constituted a distinctive array, accounting for 13 percent of the total. And the major saturated hydrocarbon identified was tricosane, rather than pentadecene (Erickson et al. 1987). *T. mammosum*'s distinctly acrid, pungent smell may reside in the linalool oxides acting in concert with tricosane. We also determined that the citruslike aroma of *T. simiarum* and *T. speciosum* flowers was due chiefly to the monoterpenoid compounds of citral, geraniol, nerol, and citronellol—a chemical profile distinct from the "oily hydrocarbon" profiles of *T. cacao* and *T. mammosum*.

As expected, *Herrania* flowers contained floral oils quite different from those of *Theobroma*. The *Herrania* floral oils were distinguished by high quantities of iridomyrmecin, guiaol, and other azulenic derivatives; these monoterpenoids and bicyclic sesquiterpenoids are absent from *Theobroma* flowers. We cannot help thinking that these distinctive compounds in *Herrania* were the principal attractants for phorid flies as pollinating agents. The occurrence of these substances in *Herrania* floral oils tends to support Cuatrecasas's (1964) systematic separation of this genus from *Theobroma* based upon morphological and other natural historical features. What is especially striking is that apparent clear differences in profiles of major constituents among the various species of *Theobroma* examined are biologically real differences that we believe influence the kinds of insects floral visitors and pollinators attracted to these species.

Given these findings, it was clear what had to be done next. The observed differences in floral oil chemistry were only as useful as their biological impact on pollination systems in the species examined. Do different kinds of *Theobroma* floral oils attract different kinds of flying insects, including known or potentially pollinating species? This became the next focus of our research at La Lola. In September 1985 the Ericksons accompanied me to La Lola for a week of fieldwork that would help set the stage for much of my subsequent field experimentation with the question of cacao pollination.

Floral Fragrances in the Field

In La Lola, we initially experimented with various kinds of commercially available insect field traps to determine which kind would be most suitable for bioassaying distilled floral fragrance oils. We settled upon the McPhail trap as the one to bioassay floral oils as insect attractants. Our tests of various entomological traps also provided us with an opportunity to survey flying insects in the La Lola cacao (Young et al. 1987b).

A McPhail trap resembles a flattened transparent glass bell, the narrow upper end open, and the lower lip at the base curved inward and upward to form a trough where water is placed. The trap works on the principal that flying insects will fly into the wide bottom opening, bounce against the glass, and then drop into the water trough and drown. A liquid chemical insect lure is suspended from a rubber stopper shoved down into the upper opening, and the entire trap is suspended in a tree by wires. McPhail traps have been used extensively in survey studies of Mediterranean fruit flies in the southern United States and California in recent years, with an appropriate pheromone or attractant used to lure the flies. The trap, using synthetic mimics of known orchid fragrance constituents, has also been used to survey orchid bees in the tropics (Bennett 1972).

Initially we set out to test subsamples of the floral oil distillates we had obtained the week before at CATIE as well as various synthetic analogues of substances known to be common insect attractants. This was a time to get acquainted with the system and how it could be useful to our project. We devised a simple way of suspending a ball of cotton from a modified paper clip shoved into the bottom side of the rubber stopper. The cotton ball then received a predetermined quantity of the substance to be tested—in this case the distilled floral fragrance. Water was added after the traps were carried into the cacao.

In a field bioassay, the fragrance disseminates into the air around the trap, establishing a concentration gradient of molecules that increases closer and closer to the cotton ball source. Flying insects that register an attraction-response to the substance through their antennae follow the concentration gradient or aerial trail leading to the source. Insects fly upwards through the large opening in the bottom of the trap towards the lure and are knocked down into the trough of soapy water, which kills them. The design of a field experiment to assess differences among various floral oils would be determined in large measure by the results of analyses of floral oil composition in the laboratory.

The floral oils obtained in September 1985 and those obtained later were divided into two portions: one for laboratory analysis, the other for field testing. We brought all the samples to Wisconsin and stored them in a freezer at the Milwaukee Public Museum until time for field testing. My plan was to conduct two or three bioassays a year, during both dry and rainy seasons. During 1986, I field tested floral oils from *T. cacao*, *T. mammosum*, *T. speciosum*, *T. simiarum*, and *H. cuatrecasana*—material distilled the previous September.

I followed the same procedure for all of the bioassays that year. We had settled on using three serial dilutions of floral oil in distilled water, namely, 1 part, 10 parts, and 100 parts per million. Using special pipettes, I made the serial dilutions in my lab area in the La Lola building and quickly dabbed the cotton balls of each trap with the appropriate amount of the test fluid. As controls, I used traps inoculated with chloroform solvent and distilled water in place of floral oils. Usually I set up eighteen experimental and six control traps and ran the bioassay for five or six consecutive days within the same cacao habitat (the cacao forest).

Using a map diagram, I assigned one or two traps to the previously tagged Matina cacao trees and added the soapy water to each trap at this time. Each trap was labeled with a number or treatment code. Usually I tested two species of floral oil at the same time, each with the same three-part dilution. On the morning of the third day of a bioassay, the traps were reinoculated in the field with remaining "fresh" oil I had stored in the refrigerator for this purpose.

Collecting insects from the twenty-four traps each morning was perhaps the most interesting part of this work. To do so, each trap was taken down from the tree and the water poured slowly through a standard coffee filter lining a small strainer, propped on the rim of a kitchen pot. I then carefully lifted the paper filter from the strainer, placed it over the stretched fingers of my left hand, and scanned the mushy gray surface with a hand lens for lodged insects, especially tiny midges. I had to look very closely since newly hatched midges in particular

One of the many McPhail traps used to determine if cacao-pollinating midges are attracted to whole, crude distilled essence or oils from cacao flowers, or to specific commercially produced hydrocarbons, which are constituents of the flower extracts. The trap is suspended from a tree branch on a wire collar and brace, which fits snugly around the top of the trap. Insects fly up through an opening in the bottom, guided by the scent from a chemical lure, hit the sides of a glass dome, and presumably tumble or drop into a water-filled trough. The water is poured through a coffee filter to catch any drowned insects. Several experiments were conducted in the La Lola cacao using these traps.

had white to translucent bodies that blended with the filter paper. Searching for midges and gently picking them off the paper with fine forceps into ethanol-filled vials was an especially trying exercise in overcast weather or during downpours. It took me about three hours to finish sampling all of the traps each morning. Emptied traps were immediately refilled with water and returned to their locations in the cacao trees.

The bioassays turned up very small numbers of biting midges and gall midges. I had a naive vision that the floral scents would pull throngs of midges, but nothing could have been further from reality. I soon considered it good hunting if one or two midges turned up in each experimental trap. But the encouraging news was an absence or much lower frequency of midges in the blanks, indicating a definite response by these insects to the floral oils and not to distilled water or the chloroform solvent. As the data began to accumulate over a two-year period,

I was pleased to learn that midges, including some species known to be cacao pollinators, were attracted to the floral oils, even though I could not detect a definitive pattern with respect to serial dilution or, to a large extent, even among the species of *Theobroma* we tested (Young et al. 1987c, 1988). The latter was somewhat surprising to me given what we had learned about the marked differences in volatile constituents among the different species. Only on rare occasions did stingless bees turn up in the traps. Most of the bees trapped, which consisted of two or three species of stingless bees and one species of halictid bee, were found in the traps baited with the floral oils from wild or noncultivated species of *Theobroma*. If there was an abundant mystery cacao pollinator lurking out there in the tropical rain forest, I was not finding it by this method. Neither did I find many other kinds of insects in the traps aside from an occasional parasitic wasp and noctuid moth and several families of Diptera, including midges, even though many other flying insects thrive in cacao (Young et al. 1987b; Young 1988b).

Because sample sizes of midges trapped using floral oil distillates or mimics of major constituents were very small, it was difficult to detect seasonal changes in the species active in and about cacao plantations. One noticeable exception was a dramatic shift in the abundance of two species of gall midges between wetter and drier periods of the year at La Lola. One species was found to be present in the drier period and absent in the wetter period; a second species exhibited the opposite pattern. No such patterns were found for biting midges, always much less numerous in the traps than gall midges, regardless of season. If gall midges are cacao pollinators, it is intriguing to consider that a seasonal replacement of key species might occur in some cacao plantations experiencing a drier period each year.

In spite of the low numbers of midges in the traps, I found this field approach helpful in one important regard. It provided me with a confirmation of something I and other cacao pollination researchers had believed all along, namely, that adult populations of pollinating midges occur at very low densities in plantations. The trapping provided a helpful measure of midge population density and how species composition and total numbers of the insects shifted at different times of the year. Even the relatively more abundant gall midges existed in very low numbers per unit area of the habitat—undoubtedly a contributing factor to observed low levels of natural pollination. The low density of adult midge populations is consistent with the general pattern of species having small populations in tropical rain forests (Slobodkin and Sanders 1969). Many diverse predatory pathways involving insects and other arthropods are characteristic of the lowland rain forest, making it a remote possibility that midge populations could build up to large densities, either in the rain forest or in the cacao.

Finding low densities of pollinators in cacao groves did not necessarily indicate a deficiency of pollinators. That usually less than 10 percent of the flowers during peak blooming periods in cacao set pods, on an annual basis, suggests that a low-density pollinator population is responsible for pollination. Consider a conservative yield goal of 350 kilograms of dry cacao beans per acre per year and assume a pod index of 26, meaning that twenty-six harvested pods produce one kilogram of dry cacao. Assume also one thousand mature cacao trees per hectare. Under conditions of the 10 percent pod-set level, there would need to be one-fourth of a pod-set per tree per day, or one pod every four days, to obtain the yield goal of 350 kilograms of dry cacao beans. A very small or low-density pollinating midge population could readily provide such a yield level—a pattern consistent with the numbers of midges obtained in the bioassays.

Fragrance Differences among Cultivars

In 1988 the distillation research took an interesting twist. I was able to obtain a vacuum steam distillation of several cultivars, or varieties, of cacao. In essence, I was taking the floral oil research a step further, from a comparison among closely related species of *Theobroma* to a finer level of analysis, that is, a comparison among several distinctive cultivars within one species. For this work Erickson lined me up with David Severson, a postdoctoral researcher working in Erickson's lab in Madison.

Before commencing a study of fragrance oils from specific cacao cultivars, I conducted a series of bioassays in 1987 and 1988 using serial dilutions of synthetic analogues of the major volatile constituents of *Theobroma* floral oils. In these studies I was able to determine that substances such as pentadecene, pentadecane, limonene, geraniol, citronellol, linalool, citral, and tricosane tested separately do not attract midges (Young 1989b). Most likely, it is the subtle interplay of several volatiles within a fragrance oil that provides the attractant. This information confirmed the importance of uncovering equally subtle differences in floral oil chemistry among different cacao cultivars and seeing how such differences might influence each cultivar's levels of attraction for pollinating midges.

In a February 1989 bioassay of floral oils in the Matina cacao forest, I was able to determine that two cultivars, Rim-100 and UF-668, attracted more biting midges and gall midges than did several others (Young 1989a). It was essentially this observation, especially since Rim-100 attracted more midges than UF-668, that prompted us to take a closer look at not only these floral oils as pollinator attractants, but also those of other cultivars. In a sense, the midges were biological detectives sorting out the most interesting floral oils from the others, making the challenge of eventually discerning the key biologically active substances in floral oils a somewhat easier task.

We wanted to distill floral oils from several cultivars or types of cacao. I selected the types based upon their range of biological characteristics, including compatibility system, floral pigmentation, and level of productivity. Included were types with well-established histories of high, intermediate, and very low pod-set. Could such differences in pod-set be related to differences in pollination levels, in part determined by differences in floral fragrance among these varieties? Were there noticeable differences in the profile of volatile constituents from floral oils among any of these cultivars, and, if so, were some cultivars more effective in attracting pollinating insects than the others? The array of cacao types I chose included some of the most familiar *criollos* and *trinitarios* in Latin America.

Between June and December 1988, Doug Dube and Paul Lynne at the Wisconsin State Laboratory of Hygiene came up with the gas chromatography and mass spectrometry data from these cultivar floral oils. In a preliminary inspection of the ion chromatograms, and by comparing many peaks with known standards, we tentatively concluded that there were distinctive differences in the profiles of the major volatile constituents among some of the cultivars.

For analysis, we compared everything to cultivar UF-613, given that its two most abundant hydrocarbons were pentadecene and pentadecane. High levels of 1-pentadecene were found in four other cultivars (SCA-6, UF-29, UF-221, and UF-668) and completely absent in Rim-100. Further analysis separated out the cultivars into two main groupings, one with relatively high levels of lower molecular weight compounds, and those with relatively low levels of these compounds. Rim-100 uniquely falls into a subgrouping having none of the lower molecular weight compounds present in its floral oil.

Thus the floral oil of Rim-100 appears to have the greatest numbers and quantity of high molecular weight compounds—perhaps an indicator of its being a more stable, effective attractant for pollinators. Interestingly, Rim-100 is phenotypically akin, in terms of floral pigmentation and pod surface texture and shape, to *criollo* cacao in Central America. Perhaps cultivars closely akin to and derived from *criollo* cacao have better floral fragrances for attracting pollinators than other cultivars. In the CATIE plantation, flowers on Rim-100 trees have been observed to support larger assemblages of aphids, ants, and other insects than is typical for most other cultivars, possibly indicating a greater attraction for insects.

This bioassay was repeated twice, once in the rainy season and once in the dry season, in the now abandoned La Tigra cacao plantation. These bioassays were designed to compare the levels of attraction of the several cacao cultivars, including Rim-100, whose floral oil compositions had been determined by GC-MS analyses. Performed during 1991, the two bioassays together netted eighty midges, of which 70 percent were trapped with Rim-100 floral oil, and of these, about 71 percent were Cecidomyiidae, including known cacao-associated species such as *Mycodiplosis ligulata*, *Aphodiplosis triangularis*, and an unidentified species of *Clinodiplosis* (Young and Severson n.d.). All four species of ceratopogonid midges trapped exclusively with the Rim-100 floral oil were either known or suspected pollinators of cultivated cacao (Young and Severson n.d.). About 83 percent of the Cecidomyiidae attracted to all floral oils tested were females. While the sample sizes are small, these results tentatively suggest that Rim-100 floral oil is a better midge attractant than those of the several other cacao cultivars being studied. Might the floral oil of Rim-100 be similar to that of truly wild cacao because it is a powerful midge attractant?

These results lead me to believe that cacao-pollinating insects might have originally coevolved with a wild variety of cacao that no longer exists or is very scarce and therefore undiscovered in the wild. Such a hypothetical ancestral variety, perhaps in some ways akin at least phenotypically to the agronomic Rim-100, may exhibit a higher level of natural pollination than most varieties cultivated today. Our preliminary evidence with the *criollo*-mimicking Rim-100 lends some tentative credibility to this idea. Under this contention, perhaps the original pollinator species associated with cacao still exists, although it could also be extinct. The evidence gathered to date clearly suggests that biting midges and gall midges, to a lesser degree, are pollinators of cacao and that honey bees clearly are not. Whether or not there is some other kind of midge, or a species of wild bee, that is the key pollinator with which cacao coevolved long ago remains to be explored—chiefly in wild populations of cacao in Amazonia.

Great caution must be exercised, however, in attaching too much significance to these preliminary differences, because sample sizes are so small, and many environmental factors can attribute to relatively small changes in midge numbers. Even slight eddies of air flow in the cacao grove can differentially affect attraction of midges to particular traps. Thus any interpretation of these results must be tempered with an awareness of these possible distortions of the data.

While this work was under way in Costa Rica and Madison, I embarked on a comparative morphological study of flowers from the same array of cacao cultivars used in the fragrance studies. I enlisted the assistance of Nancy Christy, from the Botany Section of the Milwaukee Public Museum, who carried out a detailed light microscopy study of the flowers I had brought back from La Lola. She measured eighteen distinctive floral characters and conducted the appropriate statistical analyses to reveal possible groupings of cultivars with similar characteristics.

Analyzing the floral morphology data, we showed that the cultivars separated into the same three groupings observed in the floral fragrance analyses. Rim-100 remained in a category all by itself. The interconnections, if any, between these two different kinds of data sets—floral fragrance chemistry and floral morphology at the cultivar level in cacao—remains unclear at this time. Certainly both floral fragrance and floral morphology influence and shape the interaction of the flowers with its pollinators.

These floral fragrance oil analyses of a representative albeit small sample of widely used cacao cultivars, together with previous (Young et al. 1984) and ongoing studies of *Theobroma* floral morphology, have prompted the formulation of a tentative hypothesis about altered versus natural states of pollination systems in cacao. It is my conviction at this time that the long history of artificial selection of many cultivars or horticultural races of cacao by asexual means such as cloning has altered, to varying degrees, the original or wild-type pollinator-attraction system in cacao. Under this scenario, cultivars resembling more primitive forms of cacao, such as *criollo* in the case of Rim-100, exhibit a relatively more effective, albeit still altered capacity to attract flying insects, including opportunistic pollinators such as certain dipterans, more so than cacao cultivars less akin to wild or primitive forms of cacao.

Coupled to this working hypotheses, which serves as a basis for future research, is the idea that as a result of prolonged asexual propagation of cacao cultivars for agronomic purposes, natural selection for maintaining an adequately functioning sexual breeding system in cacao has been relaxed. This condition, together with the plantation style of growing cultivated forms of cacao for commercial purposes, can help explain the generally low levels of natural pollination typical of these plantations. Under such conditions, the best that can be expected in cacao plantations, which usually contain a mixture of several artificially selected cultivars, is opportunistic, low-level pollination performed by dipterans rather than a highly specialized pollinator such as some kind or kinds of wild bee.

Even with cacao cultivars somewhat akin to a wild-type cacao, as in the case of Rim-100, there exists a somewhat dysfunctional pollinator-attraction system. We do not know the floral properties associated with this alleged dysfunctioning, manifested as ineffective or suboptimal floral fragrances. The observed reduction or structural changes in the basal ring of glandular projections observed in the flowers of cultivated cacao (Young et al. 1984) may provide some clues. As already discussed, this basal ring structure is well developed in other species of *Theobroma*. If this structure is related to the production of floral fragrances or some other component of pollinator-attraction, we might expect to find a considerable range of variation in its development among different cacao cultivars. This remains to be studied.

If cacao cultivars in plantations today represent an ecological or evolutionary anachronism in the sense that they have lost their original relationship with specialized pollinators and the ability to attract them, whatever floral attraction and reward properties are still present are being exploited by opportunistic dipterans, especially ceratopogonid and cecidomyiid midges, and occasionally stingless bees. This assertion assumes that cultivated forms of cacao have been derived from an ancestral, wild species that exhibited floral adaptations for long-distance pollination mediated by strong fliers such as wild bees. Wild or noncultivated species of *Theobroma* tend to occur as scattered, isolated trees over large areas of tropical rain forest, suggesting a general adaptation in these species for long-distance pollination to maintain genetic variation in their populations. This pollinator relationship maintained an adequate level of sexual reproduction in wild cacao, perhaps counterbalancing naturally occurring asexual propagation resulting from the growth of chupons into new trees (Chapter 6).

The examination of truly wild cacao, assuming it exists, may offer important clues to the original breeding system in this tree species, compared to what appears to be an anachronism in cultivated forms of cacao. Wild cacao on the eastern lower slopes of the Ecuadorian Andes consists of tall, subcanopy forest trees with flowers produced primarily near the tops of the trees, rather than much lower on the trunks as seen in cultivated cacao in well-maintained plantations (J. B. Allen, personal communication). Flowering near the canopy of the forest is more conducive to bee pollination than flowering confined to the more shaded depths of the forest. However, many genera and species of bees are active in the lower strata of neotropical rain forest (Roubik 1993). Stingless bees in Costa Rica exhibit a distinct preference for visiting cacao flowers in sunlight over those in shade (Young 1985b).

Although it has been noted that stingless bees rob pollen from cultivated cacao without pollinating the flowers (Soria 1975; Young 1981), the interaction of larger species of these bees and others observed visiting cacao and other *Theobroma* flowers, including halictids and euglossines, remains to be studied (A. M. Young, unpublished observations). Various wild bees are more frequent visitors to the larger, showy, and more fragrant flowers of several species of noncultivated

Theobroma (Aguiar-Falcao and Lieras 1983; A. M. Young, unpublished observations) than they are to the less conspicuous and weakly scented flowers of cultivated cacao in plantation settings. The basic floral design shared by all species of *Theobroma* (Cuatrecasas 1964) suggests a specialized adaptation for a highly coevolved pollinator relationship.

Interesting in this context is the presence in the exocrine gland secretions of various groups of wild bees of several major volatile constituents revealed by GC-MS analyses in the distilled floral fragrance oils of *Theobroma* species (Duffield et al. 1984; Erickson et al. 1987; Young and Severson, unpublished manuscript). Whether or not some wild bees obtain exocrine gland volatiles, or their precursors, from floral sources such as *Theobroma* remains to be studied and offers intriguing possibilities. Various wild bees use exocrine gland volatiles such as nerol, geraniol, citral, citronellol, limonene, and pentadecane in the production of specific airborne signals needed for mating, territorial defense, and foraging (Duffield et al. 1984). If some wild bees obtain these compounds from *Theobroma* flowers, this behavior might constitute the primary floral reward in pollination, perhaps somewhat analogous to the chemical prospecting of orchid flowers and other fragrance sources by euglossine bees in the neotropics (Williams and Whitten 1983). While it is possible that some bees might collect these chemicals from *Theobroma* flowers while pollen-thieving, others might be effective pollinators. The interaction of stingless bees with cultivated cacao stands (Soria 1975; Young 1981) might be part of the anachronism that exists today between cultivated cacao and its floral insect visitors. Whether or not cacao-pollinating midges utilize fragrance substances from *Theobroma* flowers also remains to be studied. They are clearly attracted to traps baited with synthetic analogues of some of the major volatile constituents of these floral oils (Young 1989b). The absence of bees in these traps, and their occurrence, albeit in low numbers, in traps baited with whole floral oils (Young et al. 1987b; Young 1989a) suggests that bees require a mixture of volatiles as attractant cues.

What these findings with cultivated cacao compel me to consider is that somehow, these largely asexually produced forms of cacao have lost, during their extensive and complex history of cultivation, those floral properties present in wild populations that evolved to ensure some level of long-distance pollination in natural habitats (Chapter 6). By examining floral features and insect visitors associated with noncultivated or wild species of *Theobroma,* such a perspective on the plight of cultivated cacao, in terms of poor pollination, becomes more plausible.

It is intriguing for me to speculate how my collaborative discoveries through cacao pollination research might provide further insight into cacao as a tree species in the wild. For example, I believe that the high levels of fruit-set in the abandoned Matina cacao, along with the low ratios of flowers to newly set pods (Young 1986d), are clues to the pollination biology and breeding systems of truly wild cacao populations in the rain forests of South America. These old Matina cacao trees bear very much the physique and flowering behavior of wild cacao trees. The abandoned Matina cacao, with its weak resemblance to regenerating tropical rain forest, or the older cacao trees bordering the edge of rain forest at La Tigra in Sarapiquí, compel me to think of this rain forest tree as conservative in its reproductive strategy. Its modest level of staggered flowering throughout the year, coupled with a high frequency of pod-set relative to flowering, with pods requiring six months to mature, and a highly specialized seed-dispersal mechanism all suggest an organism that, in its wild state, is well-integrated into the complex rain forest ecosystem.

Cacao does not appear at all to be a prolific maverick species of forest edges and disturbed habitats. It does not appear to be genetically programmed to fill empty natural niches quickly. Consequently, cacao is a poor contender for a commercial crop and the associated man-made niche it has been forced to occupy, the plantation habitat. This lack of suitability as a plantation crop helps explain the paucity of natural pollination that has plagued the commercialization of cacao since its plantation-type cultivation initiated by the Spaniards and British centuries ago.

Given what I have seen in Sarapiquí and La Lola, I believe that natural pollination is expedited by flying insects well-adapted to the rain forest understory and not to the open spaces of well-groomed cacao plantations. From the standpoint of the insect pollinator populations, the plantation arrangement of the cacao trees and their induced prolific flowering are unnatural attributes—and ones to which pollinating insects cannot readily respond unless coaxed into doing so by the field biologist. Inside the rain forest or in parcels of abandoned cacao, there are many niches for the breeding of cacao pollinators, but these are markedly wanting in the well-maintained cacao plantation.

Given these conditions, there is no sustainable way, save for a rethinking of how cacao is cultivated on a commercial scale, that natural pollination can be greatly increased. Each and every plantation will have its unique peculiarities when it comes to defining just how to induce the ecological expansion of pollinators into the cacao, which can result in increased pod-set, especially when the trees are healthy. One must view the current state of affairs about cacao pollination as the challenge of trying to overcome what might be called "ecological illogic": Because of cultivation practices in place for several hundreds of years, an imbalance exists between the flowering behavior of plantation cacao trees, the spatial arrangement of these trees, and the natural behaviors of pollinating insects. The behavior of pollinating insects has been molded not by selection pressures in cacao plantations but by the much more ancient evolutionary theatrics that fine-tune the multifaceted, often specialized adaptations of these organisms of the tropical rain forest.

Back to the Rain Forest

A Bridge between Agriculture and Conservation

Once, late in the day several years ago, a foreman named Moya at the La Lola research station took me on a brief journey through the cacao, one that heightened my awareness of Mesoamerica's ancient past and of the Indians who once lived on this plain. Moya knew exactly where he was going through the cacao, even though he was not following a path. As we walked along through the shade, our feet brushing through the thick layer of fallen leaves, Moya would suddenly turn, first one way and then another, and then still another. After about twenty minutes, we came to a tiny clearing along the edge of the cacao up against the rain forest. Dusk was descending fast. At one side of this clearing, inside a dense thicket of vines and small trees, Moya poked around with his machete. Hidden beneath the bramble and tangle of vines was a large rock. Moya did not cut away the cocoon of vegetation enshrouding the rock, but simply lifted it with his machete, peeling it back to one side. There before us were the features of a jaguar carved into the rock, which came up almost to my knees. It was an unfinished piece of sculpture.

Moya explained that it had been fashioned by the Indians in ancient times, but that he was perplexed why it was never completed. Someone had started it six or seven hundred years ago, perhaps even a thousand years ago. As a gentle rain began to fall, Moya covered up the rock again with the vines and led me back through the cacao. I thanked him profusely for sharing with me his secret in this spot where the cacao grove meets the rain forest, for heightening my awareness of the bond between nature and the people who inhabited this land centuries ago. The events that day crystallized for me an appreciation of the ancient peoples who first recognized the gifts of cacao. To some degree that little trek into the brush with Moya encapsulates this book—a blending of people, cacao, and tropical rain forest.

The journey of cacao through space and time reveals the continuities that exist among people in different parts of the world. La Lola is not far from the Sixaola River Valley, where *criollo* was cultivated by the Indians in ancient times. Just walking across this terrain can be a difficult matter, even today. But what must it have been like for the Indians, and later the Spaniards who explored this rugged terrain, blanketed in tropical rain forest, filled with snakes and jaguars? This land was conquered by interlopers from afar in the name of the Redeemer and Christianity, and for an embryonic system of world commerce. Yet such considerations illuminate the continuity of human existence and nature throughout the world. Today's American tropics are, to some degree, a blend of exotics, plant and animal species brought in from other parts of the world, and which generally live as biological outcasts on the fringes of nature—in gardens and marginal habitats devoid of endemic species. So too with human cultures. Central America today is a blend of the indigenous people with Europeans, Asians, Africans, and North Americans.

I sense such continuities even as I stand in my vegetable garden when August's heat scorches the Wisconsin landscape. In my garden in summer there have been squash, tomatoes, peppers, and corn—three thousand miles due north of the land about which I write, Central America. These familiar garden vegetables have their origins in Central and South America. Then, when winter's jacket of merciless cold permeates everything, I can retreat inside for a drink of hot chocolate, cinnamon, and vanilla. Mine is a sweetened concoction, not the bitter beverage from sixteenth-century Mesoamerica. Yet in such things I feel a continuity with the New World tropics. The species and good flavors we appreciate today, such as chocolate and vanilla, represent the ingenuity and basic curiosity of ancient peoples in the Americas. The results of their past tinkering persist as part of the present.

We appreciate, too, humankind's own journey and acknowledge the sophisticated prehistory of Mesoamerican peoples, and the Colonial period that ensued.

This journey intertwines nature and human history—a joining of a rugged landscape on the emerald isthmus linking two great continents, with a tilling of the soil, clearing of the forest, and plantings of coffee, bananas, and cacao. In the present we see vestiges of the past, and the present itself is the foundation, the heritage, of things yet to be. In prehistoric times, a basic curiosity about useful products in nature, whether food, medicine, or other commodities, prompted early people in the tropics and elsewhere to explore uses of various plants, including their seeds, fruits, leaves, and roots. Subsequently, Western society's discovery of these crops, already in some degree of cultivation, led to greatly expanded production in large plantations. Far north of the equator, in places like Wisconsin, life's daily routine is interjected with sips of coffee, bananas on breakfast cereal, and that special velvet taste of hot fudge on vanilla ice cream.

Behind life's routine, behind the food and pleasures embossed upon the fabric of everyday living in the highly industrialized and technological societies of today, there resides a history whose roots into our existence were forged a long time ago. This is how I view the importance of Mesoamerican prehistory, which has picked out of nature many fruits and fibers that we enjoy today. And this is the real message of the cacao tree. Much of what we enjoy out of the American tropics is not new but reaches far back into antiquity.

For these reasons and more, I have been enriched by my research on cacao pollination in Costa Rica, and I am glad to have spent considerable time in habitats where there are continuities of that splendid tropical nature I enjoy in Sarapiquí (Young 1991). What I have come to especially appreciate through my studies of cacao and its pollinating insects is the need to consider as much of the tree's natural history as possible when trying to understand the tree as an object of human intervention. The process of pollination and the factors controlling it can only be understood within the broad context of the tree's adaptations to its environment—both in natural and plantation settings.

This view of what is needed to piece together the complex puzzle of cacao pollination and natural history came to me gradually as I spent months, then years, in Costa Rica's Caribbean foothills and floodplains, exploring cacao groves and the rain forest bordering them. A sense of possible discovery permeates the inherent beauty and diversity of tropical nature. For example, just in front of a strip of *cacao abandonado* along the entrance road into La Lola, there is a well-manicured parcel of cacao trees, their branches richly dripping in delicate, clinging vines and many small epiphytes. Here, as throughout much of the La Lola cacao, rain forest trees of a mammoth girth still dot the landscape, rising sharply above the carpet of low-slung reddish and green cacao foliage. These living symbols of the rain forest, which once covered these lands long before the arrival of

Matina cacao in the sixteenth century, are festooned with huge, ropelike vines. For the most part, I can only imagine what kinds of animal life dwell in the leafy, tangled boughs of these trees. Even in their stark isolation, with the rest of the rain forest stripped away, these giant trees do not easily give up the secrets of the creatures hidden in them. Yet this too is part of the "cacao experience."

I have seen big colorful toucans and parrots, occupants of the rain forest canopy, wrestling in the branches of these big trees. Both in La Lola and La Tigra cacao, I have been surprised to flush out of the trees several keel-billed toucans. It has been especially surprising at La Tigra, since plenty of rain forest is still close at hand. Why would such wild creatures of the canopy come down into the cacao, so close to the ground, and perhaps expose themselves to unfamiliar dangers? Do toucans poke their feather-weight bills into the large, fresh, gaping holes made in cacao pods by squirrels and monkeys to pluck out any remaining seeds and eat that sweet-tasting pulp? The big Amazonian parrot might have the capability, with its powerful beak, to gnaw a hole into a cacao pod to feed on the pulp, but the toucan seems less likely to do so. At La Lola, it is in the maintained cacao and giant rain forest trees, next to the cacao forest, where I have noticed toucans on several occasions.

Here, too, I once came across a sloth crawling along the ground in the cacao. I never saw a sloth in a cacao tree, just on the ground in the cacao grove. Perhaps this is the only way the sloth can get around where the giant rain forest trees are scattered about, by coming down to the ground and crawling to the next closest big tree. It is true, of course, that sloths come down to the forest floor to defecate, so perhaps this is why I encountered one in the cacao. But seeing the sloth in the cacao, its green-tinged gray hairs glistening in the sunlight streaking through the cacao foliage, with little brown moths twirling their wings as they crawled over the poky creature's face, reminded me of an earlier scene I had witnessed at La Tigra in Sarapiquí. I had encountered a large sloth ambling slowly across a muddy road along the edge of the cacao. The animal moved with difficulty over the deep grooves created by tractor tires in the thick red mud. On one side of the road the rain forest had been recently cut down, probably compelling this sloth to move into the cacao grove.

No matter exactly where I am in the cacao within Costa Rica's sprawling Atlantic foothills and abutting lowlands, I am usually not out of eye contact with some semblance of natural forest in the distance. Nature's tenaciousness breathes through the heavy stagnant air of these lowlands when I come to visit La Lola. The ancient rain forest trees still scattered across the windswept banana fields and old cacao plantations signal nature's enduring but tenuous presence. And pockets of dense jungle growth have sprouted in places where the crops have

Fruits of Matina cacao. This tree is in a grove of abandoned Matina cacao trees bordering the La Lola Experimental Farm in Costa Rica. Matina was a kind of cacao farmed originally in Costa Rica by the Spaniards in precolonial times.

been abandoned. This rim of dense greenery occurs frequently along the russet borders of cacao groves, so evident near La Lola as well as in Sarapiquí.

An array of insects thrive in these dense stands of border vegetation up against the cacao. In some places it is not possible to reach the plantation without inching through the thick brush of the resplendent jungle. In these glades, wildly rich in plant life encroaching on the plots of cacao, bananas, and other crops, butterflies flit through the steamy air on sunlit mornings. Many organisms found in these border strips of jungle are also residents of the cacao and bananas. In some cases, their life cycles are tied to agriculture as much as to the natural vegetation cover that has been pushed aside. Clearing the land for cacao and other crops in Costa Rica's hot humid lowlands along the Atlantic has established a biological theater often quite different from that of the untouched rain forest.

Nature at these borders between rain forest and cacao shows its elusive hand in subtle, sudden encounters. Even common species may not be seen very often, depending upon many factors. A good example is the owl butterfly, *Caligo memnon*, one of the biggest butterflies in the neotropics. Although a common insect

The regenerating tropical rain forest vegetation within a light-gap in the abandoned Matina cacao grove bordering La Lola as it appeared in the late 1980s. The few cacao trees in this picture are barely discernible, covered or blocked from view by the dense growth of largely herbaceous plants and treelets. The thin, dark trunk of one Matina cacao tree can be seen in the upper left corner of the picture.

in the Atlantic lowlands of Costa Rica, it is a secretive creature in the adult stage. It is mostly active at dawn and dusk. The life of *Caligo* is closely associated with cacao groves throughout much of Central and South America. Over the years, this stately, elusive butterfly has come to symbolize for me a bridge between the cacao tree, the methods by which it is often cultivated with bananas, and the ecology of tropical rain forests. Perhaps only incidental in the biology of cacao, *Caligo* nonetheless is the key to reminding me about the human natural heritages of the neotropics.

In my experiences in Costa Rica, I have not found owl butterflies to be abundant locally. I seldom see more than a couple of individuals in the cacao or in the margin of the rain forest. At La Tigra, I particularly appreciate the connectedness of the cacao *finca* with the rain forest looming beyond. As I press my way into folds of rain forest here at the edge of the cacao, life changes swiftly. But seeing the stately owl butterfly flitting through the cacao reassures me that rain forest life spills over into the cacao. Sometimes inside the rain forest, especially in or near light-gaps, I see the owl butterfly again.

Here and there, scattered throughout the cacao at La Tigra, as at La Lola and other cacao *fincas*, there are little clumps or stands of bananas—food plants of

Caligo, the owl butterfly, perched on the trunk of a cacao tree at La Lola. This large, impressive insect lays its eggs on banana plants and their close cousins, and the caterpillars eat the leaves of these plants. Bananas are often cultivated with cacao in Costa Rica and elsewhere in Central America.

Caligo caterpillars, which emerge from eggs the butterfly has laid on the banana plants (Harrison 1963; Malo 1961; Malo and Willis 1961; Condie 1976; Young and Muyshondt 1985). Bananas are often planted as a temporary shade for young cacao trees. Just a short distance away, stands of a banana cousin, *Heliconia pogonantha*, flourish in the brushy margin of the rain forest. There is plenty of caterpillar food, yet neither the banana plants, reaching to 4 meters above the ground, nor the even taller *Heliconia* are ever defoliated here by *Caligo* caterpillars.

Various kinds of parasitic wasps and flies have also been recorded from *Caligo* eggs and caterpillars (e.g., Malo 1961). Together with other natural enemies, these parasites threaten the owl butterfly's survival. Yet in the Atlantic lowlands of Costa Rica, *Caligo* has been reported in the past as a "pest" of banana plantations (Harrison 1963). Large-scale efforts have been set into action by banana companies to control outbreaks of C. *memnon* and C. *eurilochus* (Bullock 1959). Ironically, chemical spraying to eradicate *Caligo* caterpillars in commercial banana plantations increases the chance of severe outbreaks, since spraying also kills off the parasitic insects that use these caterpillars as hosts. In the rain forests, a natural assemblage of parasites exploits *Caligo* caterpillars; this, the best form of biological control, can also serve the needs of the banana grower. When banana fields are left unsprayed, the assemblage of parasitic species likely builds up on the caterpillar population, greatly lowering the abundance of the butterfly.

It is difficult to study the owl butterfly's connections to the rain forests today. A strong flier, its natural history today is tied not only to the forest, but to the agricultural habitats that predominate. One cannot reach an understanding or an appreciation of *Caligo*'s ties to this land without being reminded of the history that brought bananas to Costa Rica in the first place. Let the owl butterfly, as it dodges and weaves its way through the cacao, flushed out of its resting spot by wandering biologists, be an impetus to recognizing the interconnections among bananas, coffee, cacao, people, and butterflies.

In many places, rain forest and cacao groves blend together, but with discernible boundaries. Modest-sized plots of lower-elevation rain forest crops abut expansive perimeters of tropical rain forest. Also, small, scattered plantings of cacao are embraced by the surrounding rain forest, which provides this crop tree with resources it needs to survive, such as assemblages of pollinating insects and an abundant flow of fallen leaves and other debris to enrich the soil as mulch. Relatively small, compact wedges of crops interspersed with rain forest, which I am proposing here, may provide a window of dual economic incentive, a marriage between forest-related agriculture and perhaps agroforestry, with conservation of what pristine and regenerating rain forest remains. There has to be a logical bridge, a balanced perspective, between economic incentives and biological conservation, a linkage between protecting the rain forest and managing already existing swatches of crops.

In its wild state, like other trees of the tropical rain forest, the cacao tree is fed by the cycle of other life. This intertwined existence among species is seen in the connections between giant trees of the canopy, and in the dense, wetted mulch of the forest floor. These two strata, far apart in the vertical plane, are close together biologically, with cacao and other subcanopy trees wedged in between them. The giant trees, dating from a few to several centuries in age, support legions of other plant life—the epiphytes. Some canopy trees are more prone to harboring these biological hitchhikers than others. The canopy layer buffers the rain forest below it from the full brunt of the intense tropical sun and rain. Through the canopy, the rain forest dissipates vast quantities of evapotranspirated water and gases into the air. It is also where greatest amount of photosynthetic activity takes place, setting the stage for the high productivity of the rain forest.

Within the subcanopy layers of the forest, trees such as wild cacao and other *Theobromas* are adapted to the densely shaded, moist conditions, which are perhaps less favorable for high rates of photosynthetic activity than the exposed primary canopy. These species nonetheless require reliable fonts of energy to survive and reproduce. The life cycle of a subcanopy-evolved species such as cacao in its wild state is closely tied to the floor or mulch cover of the rain forest, and to the upper reaches of the soil layer beneath this carpet of rot and decay. These substrates are fed, in large degree, not only by tree falls and the decay of wood, but by a rain of dead branches, sloughed-off epiphytes—sometimes whole branches covered with them come crashing down—seeds, fruits, and the carcasses of insects and other forms of animal life. The canopy of the rain forest, and the vast legions of life it supports, is eventually recycled into the mulch and soil layers on the forest floor.

From an agronomic standpoint, we are compelled to view the challenge of improved cacao production, including enhanced natural pollination and disease reduction, in terms of possible ecological discord between the tree as a rain forest organism and as a plantation cultigen. The ancient Mayas harvested cacao as essentially a rain forest tree in garden-style plantings, not as a broadly cultivated species. Cultivated cacao was most likely found in small scattered clumps among Indian settlements, similar to its arrangement in places in the wild.

Clearing the rain forest to replace or convert such small clumps of cacao trees into large plantations with neat rows of many trees necessitates a tree species well suited to crowded conditions over large areas. The cacao tree, from all available evidence, is ecologically the opposite. It is a tree of the lower strata of the rain forest, where it reproduces very conservatively. Although it does reproduce sexually by producing seed, in its wild state the cacao tree probably partitions only a relatively small amount of its energy budget into sexual reproduction (flowers, seeds, and pods) and considerably more into vegetative asexual propagation within clumps. Clumps of wild cacao are established by asexual means, with chupons growing from bent-over or fallen tree trunks and eventually becoming new trees. The wider girth and taller canopies of wild cacao trees also reflect a greater allocation of resources to vegetative growth than to sexual reproduction. Yet even under these conditions, some level of outcrossing, maintained by the long-distance transfer of pollen among different clumps of wild cacao trees, is expected to occur.

Wild cacao generally occurs as clumps near streams, habitats frequented by breeding populations of pollinating midges. Small breeding pockets of midges, with enough insects present throughout the year to pollinate a substantial proportion of available flowers in each clump, are likely associated with these scattered patches of wild cacao. Under these conditions, midges are unlikely to become satiated quickly (assuming that the female midges obtain some essential resource from cacao flowers), since the resource available on a per-flower basis is low, and there are not typically a great number of flowers at any one time.

Pollination in plantations generally occurs most frequently between neighboring trees, although successful self-pollination on self-compatible trees can be quite high (Yamida 1991). Even though pollination events between trees 9 to 12 meters apart have been observed (Yamida 1991), overall observations suggest low dispersal tendencies for cacao-pollinating midges in plantations. Whether similar patterns of natural pollination hold for wild cacao trees remains to be studied, although it is assumed that a similar strategy of natural pollination holds in the wild. Yet wild cacao, in order to survive and adapt to natural changes in its rain forest habitat, requires some level of long-distance pollen transfer, or outcrossing, mediated perhaps by larger and stronger flying insects such as wild bees (Chapter 5).

Clusters of scattered clumps of cacao trees within a relatively small area may allow for a weak to moderate level of cross-pollination, but not between clumped or individual flowering trees with greater distances between them. A spatial arrangement of isolated individual and clumped trees promotes evolutionary diversification among these different subunits of the population. Furthermore, because of the relatively small population sizes of these pockets of cacao trees, over time they would be subject to considerable genetic drift, resulting in the fixation, or accumulation, of specific sets of genes in different pockets (a phenomenon known as the Sewall Wright Effect). Such gene fixation would promote the expression of distinctive phenotypic variants, or varieties, as indeed has been described for different wild populations of cacao in Amazonia. At the same time, it is difficult to imagine that even a subcanopy tree species such as wild *T. cacao* would not require some level of genetic variation to be maintained in its population as a source of adaptation to changing conditions within its natural habitat. Outcrossing is commonly found in many tropical rain forest tree species (e.g., Bawa and Beach 1981). Thus it is expected that even within a tree species having a diversified set of naturally occurring varieties, there would be some genetic exchange between different subunits of the population, mediated by long-distance pollinators such as wild bees.

Gene drift is promoted by the isolation of small, subdivided, sexually reproducing populations and, in the case of tree species requiring animal-mediated pollination, the inability of pollinators to move about over great distances to cross-pollinate these subpopulations, or "demes." Even cross-pollination within a deme (breeding population) will result in an eventual fixation of genes and the expression of a localized phenotypic variant, oftentimes appearing quite different from other variants of the same species within a geographical region. Spontaneous mutations, together with genetic heterogeneity resulting from cross-breeding between different variants, would create a foundation of genetic material on which the combined effects of drift and local selection pressures would operate through time.

A low reward per flower, as seems to be the case in cacao, would still be sufficient to ensure pollination over such short distances, barring any effects of self-incompatibility. But in the plantation setting, midges can become satiated more quickly since many more flowers are present on a per-tree basis and certainly from tree to tree over relatively small distances. Thus in the plantation relatively few flowers are visited and pollinated compared with the total number of flowers available at any one time. Under such conditions, high levels of pollination, especially when the abundance of the pollinating insects is low, is virtually impossible to attain through natural means. My field data on the abundance of midge larvae and pupae in rotting pieces of banana pseudostems in plantations clearly point to a highly uneven, disjunct spatial distribution of midge breeding, indicating that adult midge populations in plantations most likely occur as patches, even though adults are highly mobile.

Very little, however, is known about the dispersal tendencies of cacao-pollinating midges in either rain forest or plantation habitats. Dispersal is clearly an important factor influencing the availability of pollinators, whether in reference to clumps of wild cacao or in large monoculture plantations. Perhaps in the wild the average distances among clumps of wild cacao trees exceed the dispersal capabilities of pollinators from their breeding substrates. If cacao-pollinating midges have only weak dispersal abilities, there would be little or no opportunity for the transfer of pollen between different, widely separated clumps of wild cacao in the rain forest. Each cacao clump would be essentially an evolutionary island, a genetic dead end or bottleneck. Interpatch movement of pollen would be minimal, cutting off the population from sustaining some level of genetic variation (except for that arising through somatic mutations). Yet the fairly uniform environment of the forest understory may preclude a high selective premium being placed on maintaining high levels of genetic variation in wild cacao.

Such a scenario could only work if truly wild cacao was genetically self-compatible, rather than self-incompatible, the more widely held view. Self-incompatibility in plant breeding systems is most likely to evolve under environmental conditions in which pollinators are abundant, thus facilitating the transfer of pollen over separate, scattered plants or among closely juxtaposed plants. Self-compatibility is more adaptive when pollinators are rare or absent, or when their behavior is unpredictable, short-circuiting the plant's dependency upon these animals for successful cross-fertilization.

A scattered arrangement of trees appears best suited for a bee-mediated pollination system, since wild bees are strong fliers over large distances and can carry large quantities of pollen on their hairy bodies. Yet all data thus far from pollination studies in plantation cacao indicate an absence of bees as pollinators. This may not necessarily be the case in truly wild cacao in rain forest habitats. There is a pitiful dearth of field observations on floral visitors to wild cacao. In the tropics many kinds of bees obtain essential resources other than nectar and pollen from certain kinds of flowers, including fragrance oils and nutritive lipids (e.g., Simpson and Neff 1981). Because bees provision nests filled with their young and therefore have higher energy demands than midges, they are considered to be more ubiquitous pollinators than Diptera (midges and flies). Yet some Diptera may also have highly specialized resource requirements, such as fragrance or resin oils, similar to some tropical bees.

The intrinsic biology of bees requires the availability and exploitation of constant, predictable sources of food, usually incorporating several different kinds (i.e., the flower rewards of several different kinds of plants in a habitat), even though the relative abundance, and therefore bee dependency, of these resources may fluctuate dramatically throughout the year. This is much less likely to be the case for midges as pollinators because adult midges do not take care of their offspring.

Although bee species may have highly patchy distributions in rain forests, especially along stream edges, light-gaps, and margins, their high dispersal tendencies allow them to visit flowers over large areas. Midges may not be able to disperse over large areas as well as bees. If this is true, low dispersal tendencies in midges, coupled with low reproductive levels, would result in low-density, highly localized pollinator pools in rain forests, and certainly even more so in large-scale monoculture plantation settings. Because cacao flowers drop off the tree if they are not pollinated within a relatively short time (about forty-eight hours), and because they present few resource rewards per flower and have a pollen viability of less than a half-day, a low-density, patchy pollinator population cannot adequately visit a large portion of the flowers available in an entire plantation in a single day. Thus, quite expectedly, natural pollination rates are very low under these conditions. But truly wild cacao might possess a more effective pollinator attraction system than cultivated forms or horticultural races, ensuring adequate pollination by a highly specialized pollinator such as bees (Chapter 5).

Given the above considerations about wild cacao, it is highly impractical to redesign cacao plantations to maximize pollination by a specialized group of pollinators such as wild bees. Wild bees are not attracted in great numbers to the flowers of cacao cultivars in plantation settings. What little is known about the interaction of bees with the flowers of cacao cultivars indicates that these insects do not necessarily pollinate the flowers. Under certain conditions wild bees such as halictids may be important pollinators of cacao in plantations (Kaufmann 1975). Because the chocolate industry depends upon the agronomic performance of cacao cultivars in plantations, any efforts to increase pollination by insects under these conditions ought to focus on improved means of managing midge populations as opportunistic pollinators of plantation cacao trees. It is impractical to grow cacao as scattered, isolated small clumps to mimic the tree's natural distribution in tropical rain forest. We know very little about agronomic parameters such as pod yield and bean quality in wild cacao. Therefore, enhancing yields in plantations must focus upon increasing the pollination of flowers by midges on cacao cultivars.

The implication of these considerations of midge biology is that midges might be adequately abundant in wild habitats in the American tropics, but they are scarcely abundant in managed, large-scale plantations. The ancient Mayas may have achieved substantial pollination by keeping their stands of cacao small, in the manner of gardens. Such an arrangement, especially within close proximity to actual rain forest, may provide sufficient numbers of pollinators to ensure a modest annual crop of cacao pods. Based on studies of midge populations, cacao pollinators may be encouraged to intensify their breeding in cacao plantations by increasing the availability of favored breeding substrates. The distribution and abundance of breeding sites in managed or commercial cacao plantations throughout the tropics is a limiting factor in the abundance of cacao-pollinating midges. Breeding sites for midges may also be limited in natural forests, but they are much more so in well-groomed cacao plantations.

Compared with cover provided by the tropical rain forest, the relatively reduced shade cover typical of many cacao plantations, while promoting plant growth and flowering, places stress on the breeding of pollinator midge populations. Under the rain forest canopy, midges, owing to their small body size and breeding in wet debris, presumably thrive in ecologically diverse pockets of high humidity—suitable conditions for maintaining the moist plant debris needed for breeding and the survival of adult midges. Layers of high humidity near the ground cover are much more disjunct in many cacao plantations because of the reduced shade cover—an agronomic Catch-22 in which reduced shade favors more flowering, quicker tree growth, and lowered fungus attack on developing pods, while stressing pollinator populations at the same time. Under these conditions, in which shade is moderate, low, or even absent, breeding sites dry up and become useless for midges, and midge populations dwindle, attaining levels of abundance well below the carrying capacity of resource conditions found in tropical rain forest. Plantation pollinator populations, left to their own accord, cannot attain, under most circumstances, densities or abundance levels approaching the potential of natural rain forest habitats.

Such negative pressures on pollinators can be alleviated in part by rethinking the design of the cacao plantation to encourage the stabilized proliferation of midges. There is a means to this important end without necessarily imposing a greater shade cover on the plantation. Higher levels of cacao tree flowering achieved in plantations, compared with the suppressed flowering typical in wild cacao, makes it compelling to discover ways of field-manipulating pollinator populations.

One way to render the cacao plantation more closed from the standpoint of pollinators, as shown in midge substrate studies, is to add banana pseudostems to the ground cover within the plantation habitat. Because decaying pseudostem pieces retain considerable water under the tree cover, they are ideal microcosms for maintaining breeding populations of cacao pollinators through the year, even during the tropical dry season. If it is possible to establish a stabilized population of pollinators in the plantation by such a method, then pollinators would be more or less evenly available throughout the year, in spite of swings in seasonality, to track the flowering cycles of the cacao trees. This would circumvent the commonly encountered asynchrony between periods of high flowering in cacao and later periods of high abundance of pollinating midges needed to produce the cacao crop.

Ancient Indians of Mexico and Central America devised ways to grow cacao in moist habitats within dry regions, as demonstrated by the cenotes of the ancient Mayas in the northern Yucatán peninsula and Indians of the Nicoya peninsula in Costa Rica, who grew cacao along densely shaded river banks. These techniques may have also benefited pollinator populations. What these ancient peoples created was a situation somewhat similar to the natural distribution of wild cacao in rain forests. They established a spatially patchy distribution of cultivated cacao trees with the essential proviso that these patches be highly moist assemblages of various tree species, including cacao, and in topographical features guaranteed to retain high humidity throughout the year.

In these confined structures or natural enclosures—the cenotes and steep river banks—adequate levels of midges may have been available to pollinate cacao trees. Levels of flowering might have been lower than what would be agronomically acceptable levels today, but the ratio between total flowers present and the numbers pollinated might have been more even than what now occurs in large plantations in humid tropics. A similar achievement could have been obtained by the ancient Indians at Izalco on the dry Pacific coast of El Salvador. They channeled water from the mountains into carefully designed irrigation canals. These modest cacao groves were serviced with year-round water supplies, establishing conditions necessary to encourage the breeding of midges within accumulating leaf litter and other plant debris along the irrigation ditches.

Thus it appears likely that ancient peoples living in the driest regions of Mesoamerica achieved a successful level of cacao cultivation by growing small or modest stands of the trees within isolated and scattered wet or moist pockets peculiar to the topography (cenotes and steep river banks) or in man-made structures (irrigation canals within well-shaded cacao groves). Under these conditions ancient peoples probably were able to maximize pollination and minimize fungal disease of pods, a set of conditions that pushes production upwards. Archival materials of early Spanish explorers in the sixteenth century attest to the bounty of the indigenous cacao production they found.

In large-scale modern plantations, natural pollination of cacao trees can be improved by a combination of practices. The most important ones include: maintaining borders and corridors of forest vegetation as reservoirs of pollinating insects available to move into the adjacent cacao and pollinate; maintaining adequate shade with a well-developed ground cover of leaf litter to set the initial stages for pollinators to breed in the cacao, leaving behind discarded pod husks for this purpose as well; implementing a regular program (every two or three months) to add piles of banana pseudostem pieces to the leaf litter in the cacao to enhance midge breeding within the plantation; maintaining cacao trees in a healthy state with adequate disease control and fertilizer; and exploring which combinations of cacao varieties fare best in a particular plantation to ensure adequate pollen flow or exchange between self-compatible and self-incompatible varieties.

In the cultivation of cacao for commercial purposes, it is important to identify those types suitable for production under different environmental conditions. Types must be selected on a region-by-region or locality-by-locality basis if the cacao grower expects to maximize yields. Solutions to the problem of low cacao production must be forged in accordance with the local conditions—what works in one region, be it midge breeding substrate or choice of the cacao variety most resistant to prevailing disease, may not work elsewhere. It is common knowledge that grapes grown on one side of the Rhine River in Germany yield a wine very different from that produced on the other side. Ever since the days of Sir Walter Raleigh, the British have sought to grow superior Virginia tobacco in other places, such as Rhodesia (now Zimbabwe), only to obtain inferior forms of the crop. Such findings, made repeatedly throughout the history of agriculture, bear a noteworthy message for the challenges of cacao production.

Preserving borders of natural rain forest and other types of vegetation along the perimeters of cacao plantations in modern-day Central America holds promise of improving cacao production while also protecting what little natural habitat exists today in cacao-growing regions. We must build upon the wisdom of ancient Mesoamerican peoples in this regard and recognize that maintaining the natural forest in close proximity to agricultural land does not mean attracting only pestilence and disease, but also beneficial organisms such as pollinators and symbionts within different trophic levels. Breaking up large plantations with thick "hedgerows" of wild vegetation may help to ensure a supply of pollinating insects. Little is known about the broad ecological needs of ceratopogonid and cecidomyiid midges associated with cacao (Chapter 5), and it might well be that heterogeneous assemblages of native plant species, as in natural habitats, provide a range of these resources that allow midge populations to survive. This concept can be applied to cacao farming to fuse increased economic gains from this cash crop with biological conservation of natural vegetation and the plethora of largely unstudied animal life associated with it.

In his explorations of the Venezuelan interior, Humboldt (1884) noted that newly created cacao plantations enclosed by rain forest are highly productive: "They become there the more productive, as the lands, newly cleared and surrounded by forests, are in contact with an atmosphere more damp, more stagnant and more loaded with mephitic exhalations." What is emerging from our current research in Costa Rica on the enhanced breeding of cacao-pollinating midges in rotten pieces of banana pseudostems is a similar observation. Small cacao groves, those of less than 50 acres, stand a better chance of increased productivity of cacao beans if they are enclosed by rain forest, since these forests are the sources of pollinating midge populations. The rain forest provides the "colonist" midges that breed in the banana pseudostems distributed in the cacao, more so than if there was little or no forest in the vicinity of the cacao plantation.

Quite inadvertently, my discoveries about the breeding requirements of cacao-pollinating midges and their natural history in Central American plantations heightened my curiosity not only about wild cacao, but also about the broader picture of how this tree was originally cultivated by ancient peoples on the isthmus. In one sense, my field pollination studies corroborated the wisdom of ancient peoples who adequately farmed cacao in Mesoamerica as small plots or groves, likely in close proximity to rain forest. The successful prehistoric introduction of cacao farming into drier parts of the isthmus, such as the extremely dry Nicoya Peninsula in Costa Rica or the northern Yucatán peninsula, benefited from the Indians' knowledge that cacao trees require high air humidity and a moist soil and ground cover in order to thrive—conditions also promoting natural pollination. Growing cacao in small groves within rain forest, or in year-around moist or wet places within dry zones of Mesoamerica, likely minimizes fungal disease. Disease becomes more important as the size of plantations increase.

My conviction, based on my experiences with cacao, is that prehistoric peoples in Mesoamerica had the right approach to farming cacao. Relatively small cacao orchards enclosed by forest or set within small, diversified plots of various crops within the forest optimized the chances for high productivity. Pollination of cacao trees under these conditions may have been high or at least substantially higher than what is typically found in large, monoculture-type cacao plantations today. Similar results in modern times with large plantations, according to this view, can only be achieved by costly management of pests and diseases, application of expensive chemical fertilizers, and the manipulation of pollinator populations. All of these challenges were minimized by the ingenuity of the Mayas, Aztecs, and other Mesoamerican peoples.

Tracing cacao's journey from a wild rain forest tree to a prehistoric crop to a modern global commodity has spurred my own journey as a biologist. Through direct field study, coming to understand the cacao tree, especially its life cycle and relationships with the tropical rain forest in Costa Rica, has enabled me to broaden my own experience and perspective on tropical biology—both from the basic research and applied or agronomic viewpoints. In some ways, the challenge at hand is about building bridges—first, by linking an appreciation of basic tropical biology in the rain forest with cacao agriculture, and second, by applying the goals of a biologist using basic research techniques and approaches to questions in cacao agriculture. These bridges were not built overnight or in a year or two. They came into being gradually over a decade marked with annual fieldwork in Costa Rica. Only in this manner would I come to realize that an agriculturally important tree, *Theobroma cacao,* transposed from the tropical rain forest into orchards and plantations in the Americas, is still ecologically tied to the rhythms of the rain forest.

It is possible to preserve the natural beauty and ecological integrity of the fragments of tropical rain forest still left in places like Caribbean coastal Costa Rica. We can accomplish this goal by considering the long-term benefit of these habitats on production in cacao, one of the most important crops of the entire region for the small-scale farmer. Appreciating the ecological linkages of the cacao tree to its rain forest heritage—through natural pollination and other features of its intricate natural history—provides helpful insights, and perhaps a refreshed wisdom, in protecting the tropical rain forest as an aid to economic development. Protection of endangered tropical forest habitats can be linked to attempts at natural improvement or enhancement of agronomic performance in small-to-moderate plantings of crops such as cacao.

Short of restructuring the cacao tree genetically through biotechnology, we cannot lose sight of the long-standing evolutionary blueprint that adapts the tree chiefly to the forest, not the plantation. During two thousand years of cultivation, the opportunity for artificial selection to alter the tree has certainly been available, but the bulk of its biological identity is still as a tropical rain forest tree. Given this consideration, old cacao plantations in Central America may be revitalized by appreciating the role of the tropical rain forest in supplying pools of pollinators to the tree—insects that may not be solely dependent upon this tree species, but which the tree is clearly dependent upon for reproduction. Consideration can also be given to reintroducing strip-plantings of cacao into riverbank habitats, where the species was previously cultivated by Indians within dry regions.

Because the cacao tree is still tied ecologically to the tropical rain forest, the more we can incorporate certain features of the rain forest into the design of commercial plantations, the more likely we are to maintain production at suitable economic levels in the long run. This may necessitate recognizing what the Mayas, Aztecs, and other Mesoamerican peoples realized long ago—that growing the "food of the gods" in small groves can reap considerable harvests.

Without even knowing it, these ancient peoples may have learned to cultivate cacao in a manner consistent with good natural pollination and disease control. In such modesty there can be marvelous accomplishments, namely the delightful discovery of chocolate long ago and the livelihoods of rain forest zone farmers both then and now. To this I also add the insight that comes with interpreting the cacao tree as a creature of nature, something that compels us to protect what little extant, collective tropical nature is left in the ecologically and culturally rich isthmus of Mesoamerica. In doing so, we might well be ensuring, in the long run, the world's coveted supply of chocolate, while protecting many more biological riches integrated into the complex fabric of tropical diversity.

Appendix

Names of Plants and Animals Mentioned in the Text

Common Name	Scientific Name
Plants	
Cocoa (=cacao)	*Theobroma cacao*
Wild cacao (=*cacao silvestre* or *cacao de monte*)	*Theobroma speciosum*
	Theobroma simiarum
(=*patlaxi*, or *pataxte*)	*Theobroma bicolor*
(=*cushta*)	*Theobroma angustifolia*
	Theobroma pentagona
	Theobroma grandiflorum
	Theobroma mammosum
	Herrania spp.

Chile pepper	*Capsicum* spp.
Vanilla	*Vanilla planifolia*
Cinnamon	*Cinnamomum zeylanicum*
Corn (=maize)	*Zea mays*
Wheat	*Triticum* spp.
Rice	*Oryza sativa*
Squash (=gourd)	*Curcubita* spp.
Potato	*Solanum tuberosum*
Tomato	*Lycopersicon esculentum*
Beans	*Phaseolus* spp.
Annatto (=achiote)	*Bixa orellana*
Pará rubber	*Hevea brasilensis*
Wild rubber	*Hevea guianensis* *Hevea benthamania*
Castilloa (=caucho rubber)	*Castilla* spp.
Peanuts (=*cacao de la tierra*)	*Arachis hypogea*
Cotton	*Gossypium* spp.
Coca	*Erythroxylum coca*
Tea	*Camellia sinensis*
Yuca (=cassava, manioc)	*Manihot esculenta*
Avocado	*Persea americana*
Tobacco	*Nicotiana tadacum*
Madre de cacao	*Gliricidia sepium*
Guama	*Inga* spp.
Coral tree	*Erythrina* spp.
Coconut	*Cocos nucifere*
African oil palm	*Elaeis guineensis*
Soursop (=custard apple)	*Annona* spp.
Citrus	*Citrus* spp.
Monkey ladder vine (=monkey vine or *escalera de mono*)	*Bauhinia guianensis*
Banana	*Musa* spp.
Manila hemp	*Musa textilis*
Silk fig	*Musa paradisiaca*
French plaintain	*Musa sapientum*
Wild bananas	*Musa acuminata* *Musa balbisiana*
Wild plaintain	*Heliconia* spp.
Gum tree	*Sterculia chicha*
Kola	*Cola* spp.
Guazuma	*Guazuma* spp.
Laurel tree	*Cordia alliodora*
Manu tree	*Minquartia guianensis*
Monkey pot tree	*Lecythis costaricensis*
Siparuna tree	*Siparuna* spp.
Goethalsia tree	*Goethalsia meiantha*

Trema tree	*Trema* spp.
Trumpet tree	*Cecropia* spp.
Dutchman's pipe vine	*Aristolochia* spp.
Maxillaria orchid	*Maxillaria* spp.

Fungi

Panama disease (=*mal de Panamá*)	*Fusarium cabense*
Witches-broom	*Crinipellis perniciosus*
Coffee rust	*Hemileia vastatrix*
Monilia	*Moniliopthora roreri*
Black pod	*Phytophthora* spp.

Insects

Owl butterfly	*Caligo* spp.
Bala ant	*Paraponera clavata*
Leaf-cutter ant	*Atta* spp.
Leaf beetles	*Monolepta* spp.
	Colaspis spp.
Biting midges	*Forcipomyia* (=*Euprojoannisia*) spp.
	Dasyhelea spp.
	Atrichopogon spp.
Gall midges	*Mycodiplosis ligulata*
	Clinodiplosis
	Aphodiplosis spp.
Stingless bee	*Trigona jaty*
Africanized honey bee	*Apis mellifera scutellata*
Crematogaster ant	*Crematogaster* spp.
Prepona butterfly	*Prepona* spp.

Birds

Keel-billed toucan	*Ramphastos sulfuratus*

Reptiles

Fer-de-lance	*Bothrops asper*

Amphibians

Bufo toad	*Bufo marinus*

Mammals

Sloth (three-toed)	*Bradypus* spp.
Jaguar	*Felis onca*
White-faced monkey	*Cebus capucinus*
Agouti	*Dasyprocta punctata*
Squirrel	*Sciurus variegatoides*

Bibliography

Adams, R. E. 1990. Salvaging the past at Río Azul, Guatemala. *Terra* 29:17–26.

Aguiar-Falcao, M., and E. Lieras. 1983. Aspectos fenologicos, ecologicos e de produtividade do cupucao *(Theobroma grandiflorum)* (Willd. ex Spreng.) Schum. *Acta Amazonia* 13:725–35.

Alexander, I. J. 1986. Systematics and ecology of ectomycorrhizal legumes. In *Advances in legume biology*, ed. C. H. Stirton and J. L. Zarucchi, 067–625. Monographs in Systematic Biology of the Missouri Botanical Garden. St. Louis: Missouri Botanical Garden.

Allen, J. B. 1982. Collecting wild cocoa at its centre of diversity. In *Proceedings of the Eighth International Cocoa Research Conference, Cartagena, Colombia*, 655–62. London: Cocoa Producers' Alliance and Stephen Austin and Sons.

Allen, J. B., and R. A. Lass. 1983. London Cocoa Trade Amazon Project. Final report, Phase 1. *Cocoa Growers' Bulletin* 34:1–71.

Allen, P. H. 1956. *The rain forests of Golfo Dulce*. Gainesville, Fla.: University of Florida Press.

Alvim, P. de T. 1977. Cacao. In *Ecophysiology of tropical crops*, ed. P. de T. Alvim and T. Kozlowski, 279–313. New York and London: Academic Press.

Andrews, R. M. 1979. *Evolution of life histories: A comparison of* (Anolis) *lizards from matched island and mainland habitats*. Breviora no. 454. Cambridge, Mass.: Museum of Comparative Zoology, Harvard University.

Ayorinde, J. A. 1966. Historical notes on the introduction and development of the cocoa industry in Nigeria. *Nigeria Agricultural Journal* 3:18–23.

Balee, W. 1989. The culture of Amazonia forests. *Advances in Economic Botany* 7:1–21.

Barrau, J. 1979. Sur l'oriine du cacaoyer, *Theobroma cacao* Linne, Sterculiacees. *Journal d'Agriculture Traditionelle et de Botanique Appliquee* 26:171–80.

Bartley, B. G. D. 1964. Studies of quantitative characters: Genetic studies. *Annual Report on Cacao Research* 27:28–33. Imperial College of Tropical Agriculture, Trinidad.

———. 1968. The development of superior varieties and progeny trials. *Annual Report on Cacao Research* 31:11–22. Imperial College of Tropical Agriculture, Trinidad.

Bawa, K. S., and J. H. Beach. 1981. Evolution of sexual systems in flowering plants. *Annals of the Missouri Botanical Garden* 68:254–74.

Bennett, F. D. 1972. Baited McPhail fruitfly traps to collect euglossine bees. *Journal of the New York Entomological Society* 80:137–45.

Berdan, F. F. 1982. *The Aztecs of Central Mexico*. New York: Holt, Rinehart and Winston.

Bergmann, J. F. 1969. The distribution of cacao cultivation in pre-Columbian America. *Annals of the Association of American Geographers* 59:85–96.

Billes, D. J. 1941. Pollination of *(Theobroma cacao)* in Trinidad. *Tropical Agriculture* (Trinidad) 18:151–56.

Borheygi, S. F. 1956. Archaeological synthesis of the Guatemalan highlands. In *Archaeology of Southern Mesoamerica*, pt. 1, ed. G. R. Willey, 3–58. Vol. 2 of *Handbook of Middle American Indians*, R. Wuachope, gen. ed. Austin: University of Texas Press.

Borroughs, H., and J. R. Hunter. 1963. The effect of temperature on the growth of cacao seeds. *Proceedings of the American Society for Horticultural Science* 82:222–24.

Bullock, R. C. 1959. Leaf-feeding caterpillars and their control. *Research Newsletter* (United Fruit Company) 6:7–15.

Bystrak, P. G. and W. W. Wirth. 1978. *The North American species of* Forcipomyia, *subgenus* Euprojoannisia *(Diptera: Ceratopogonidae)*. U.S. Department of Agriculture Technical Bulletin no. 1591. Washington, D.C.: U.S. Government Printing Office.

Cadbury, R. [Historicus, pseud.]. 1896. *Cocoa: All about it*. London: Low, Marston.

Campbell, D. G., D. C. Daly, G. T. Prance, and U. N. Maciel. 1986. Quantitative ecological inventory of terra firme and varzea tropical forest on the Rio Xingu, Brazilian Amazon. *Brittonia* 38:369–93.

Chalmers, W. F. 1972. The performance of Scavina hybrids in Trinidad and their future role. In *Proceedings of the Fourth International Cocoa Research Conference, St. Augustine, Trinidad*, 99–113. Port-of-Spain: Government of Trinidad and Tobago Publication.

Cheesman, E. E. 1944. Notes on the nomenclature, classification and possible relationships of cacao populations. *Tropical Agriculture* (Trinidad) 21:144–59.

Ciudad-Real, A. de. 1872. Relación breve y verdadera de algunas cosas de las muchas que sucedieron al Padre Fray Alonso Ponce en las Provincias de la Nueva España. Vols. 57–58 of *Colección de documentos inéditos para la historia de España*. Madrid: Imprenta de la Viuda de Calero.

Clark, J. C., ed. and trans. 1938. *Codex Mendoza*. London: Waterlow & Sons.

Condie, S. 1976. Some notes on the biology and behavior of three species of Lepidoptera (Satyridae: Brassolinae) on non-economic plants in Costa Rica. *Tebiwa* 3:1–28.

Cook, L. R. 1982. Chocolate production and use. Ms. revised by E. H. Meursing.

Cope, F. W. 1939. Agents of pollination in cacao. *Annual Report on Cacao Research* (Trinidad) 1939:13–19.

———. 1958. Incompatibility in *Theobroma cacao*. *Nature* (London) 181:279.

Cuatrecasas, J. 1964. Cacao and its allies: A taxonomic revision of the genus *Theobroma*. *Contributions from the U.S. National Herbarium* 35:379–614.

Dahlin, B. 1979. Cropping cash in the postclassic: A cultural impact statement. In *Maya archaeology and ethnohistory*, ed. N. Hammon and G. Wiley. Austin: University of Texas Press.

Dessart, P. 1961. Contribution a l'etude des Ceratopogonidae (Diptera). *Bulletin Agriculturique du Congo Belge* 52:525–40.

Dodge, B. S. 1979. *It started in Eden*. New York: McGraw Hill.

Duffield, R. M., J. W. Wheeler, and G. C. Eickwort. 1984. Sociochemicals of bees. In *Chemical Ecology of Insects, ed. W. J. Bell and R. T. Carde, 387–420*. Sunderland, Mass.: Sinauer Associates.

Enríquez, G. A., and J. V. Soria. 1967. *Cacao cultivars register*. San José, Costa Rica: IICA Teaching and Research Center.

Entwistle, P. F. 1972. *Pests of cocoa*. London: Longman.

Erickson, B. J., A. M. Young, M. A. Strand, and E. H. Erickson, Jr. 1987. Pollination biology of *(Theobroma)* and *(Herrania)* (Sterculiaceae). II. Analyses of floral oils. *Insect Science and Its Application* 8:301–10.

Erickson, E. H., Jr., A. M. Young, and B. J. Erickson. 1988. Pollen collection by honey bees (Hymenoptera: Apidae) in a Costa Rican cacao *(Theobroma cacao)* plantation. *Journal of Apicultural Research* 27:190–96.

Erneholm, I. 1948. *Cacao production of South America: Historical development and present geographical distribution*. Gothenberg: Goteborgs Hogskolas Geografiska Institution.

Etter, H. 1971–73. Out of a chrysalis, a Wardian case. *Morton Arboretum Quarterly* 7–9:49–55.

Feil, J. P. 1992. Reproductive ecology of dioecious *Siparuna* (Monimiaceae) in Ecuador: A case of gall midge pollination. *Botanical Journal of the Linnaean Society* 110:171–203.

Feinsinger, P. 1983. Coevolution and pollination. In *Coevolution*, ed. D. J. Futuyma and M. Slatkin, 282–310. Sunderland, Mass.: Sinauer Associates.

Fernández, L. 1886. Provanca hecha a pedimento de Juan Vázquez de Coronado acerca de sus méritos y servicios: Año de 1563. In *Colección de documentos para la historia de Costa Rica*, vol. 4, 288. París: Imprenta P. Dupont.

Fowler, W. R., Jr. 1985. Ethnohistoric sources on the Pipil-Nicarao of Central America: A critical analysis. *Ethnohistory* 32:37–62.

———. 1987. Cacao, indigo and coffee: Cash crops in the history of El Salvador. *Research in Economic Anthropology* 8:139–67.

———. 1989a. *The cultural evolution of ancient Nahua civilizations: The Pipil-Nicarao of Central America*. Norman, Okla.: University of Oklahoma Press.

———. 1989b. *The Pipil of Pacific Guatemala and El Salvador*. In *New frontiers in the archeology of the Pacific coast of southern Mesoamerica*, ed. F. Bove and L. Heller, 229–42. Anthropological Research Papers, no. 39. Tempe: Arizona State University Press.

———. 1989c. Household differentiation and the production of wealth: The case of sixteenth century Izalco, El Salvador. In *Household differentiation: Cases from the Mesoamerican lowlands*. Symposium of the 54th Meeting of the Society for American Archaeology, organized by A. Ford and P. McAnany, Apr. 5–9, 1989, Atlanta. Preprint.

Friede, J. 1953. *Los Andakí. 1538–1947, historia de la aculturación de una tribu selvática*. México: Fondo de Cultura Económica.

Gagné, R. J. 1984. Five new species of neotropical Cecidomyiidae (Diptera) associated with cacao flowers, killing the buds of Clusiaceae, or preying on mites. *Brenesia* (Costa Rica) 22:123–38.

García de Palacio, D. 1881. Relación hecha por el Licenciado Palacio al Rey D. Felipe II, en la que describe la provincia de Guatemala, las costumbres de los indios y otras cosas notables. In *Colleción de documentos para la historia de Costa Rica*, by L. Fernández, vol. 1, 1–52. San José, Costa Rica: Imprenta Nacional.

Gehring, C. A., and T. G. Whitham. 1991. Herbivore-driven mycorrhizal mutualism in insect-susceptible pinyon pine. *Nature* (London) 353:556–57.

Gentry, A. 1982. Phytogeographic patterns as evidence for a Choco refuge. In *Biological diversification in the tropics*, ed. G. T. Prance, 12–136. New York: Columbia University Press.

Glendinning, D. R. 1962. Natural pollination of cocoa. *Nature* (London) 193:1305.

Gómez-Pompa, A., J. S. Flores, and M. A. Fernández. 1990. The sacred cacao groves of the Maya. *Latin American Antiquity* 1:247–57.

Haffer, J. 1969. Speciation in Amazonian rain forest birds. *Science* 165:131–37.

———. 1982. General aspects of the refuge theory. In *Biological diversification in the tropics*, ed. G. T. Prance, 6–24. New York: Columbia University Press.

Hall, D. W., and B. V. Brown. 1993. Pollination of *Aristolochia littoralis* (Aristolochiales: Aristolochiaceae) by males of *Megaselia* spp. (Diptera: Phoridae). *Annals of the Entomological Society of America* 86:609–13.

Hardy, F. 1935. The chemical and ecological researches on cacao. *Tropical Agriculture* (Trinidad) 12:175–78.

Harrison, J. O. 1963. On the biology of three banana pests in Costa Rica (Lepidoptera: Limacodidae, Nymphalidae). *Annals of the Entomological Society of America* 56:87–94.

Hart, J. H. 1891. *Cacao*. Port-of-Spain, Trinidad: Mirror Office.

———. 1911. *Cacao: A manual on the cultivation and curing of cacao*. London: Duckworth.

Hernández, F. 1651. *Rerum medicarum Novae Hispaniae thesaurus, seu plantarum, animalium, mineralium mexicanorum historia*. Rome: Lincei Edition.

Hernández, J. 1965. Insect pollination of cacao (*Theobroma cacao* L.) in Costa Rica. Ph.D. diss., University of Wisconsin, Madison.

Hickling, D. F. 1961. *Inland tropical fisheries*. London: Longmans.

Hinchley Hart, J. 1911. *Cacao*. London: Duckworth.

Holdridge, L. R. 1950. Notes on the native and cultivated cacaos in Central America and Mexico. *Cacao Information Bulletin* 2:1–5.

Humboldt, Alexander von. 1884. *Travels to the equinoctial regions of America, during the years 1799–1804*. Vols. 1–3. London: George Bell & Sons.

Hunter, J. R. 1961. The status of cacao variety improvement work in the Western Hemisphere. *Cacao* (Costa Rica) 6:2–16.

———. 1990. The status of cacao (*Theobroma cacao*, Sterculiaceae) improvement in the Western Hemisphere. *Economic Botany* 44:425–39.

Janos, D. P. 1980. Vesicular-arbuscular mycorrhizae affect lowland tropical rain forest plant growth. *Ecology* 61:151–62.

Janson, C. H. 1983. Adaptation of fruit morphology to dispersal agents in a neotropical forest. *Science* 219:187–89.

Janzen, D. H. 1971. Seed predation by animals. *Annual Review of Ecology and Systematics* 2:465–92.

Kaufmann, T. 1973. Preliminary observations on cecidomyiid midge and its role as a cocoa pollinator in Ghana. *Ghana Journal of Agricultural Science* 6:1093–98.

———. 1975. An efficient, new cocoa pollinator (Lasioglossum) sp. (Hymenoptera: Halictidae) in Ghana, West Africa. *Turrialba* (Costa Rica) 25:90–91.

Kevan, P. G., and H. G. Baker. 1983. Insects as flower visitors and pollinators. *Annual Review of Entomology* 28:407–53.

Lange, F. W. 1971. *Cultural history of the Sapoa Valley, Costa Rica.* Occasional Papers in Anthropology, no. 4. Beloit, Wisc.: Logan Museum of Anthropology, Beloit College.

Lara, F. E. 1957. Estudio preliminar sobre la entomología económica del cacao en la zona Atlántica de Costa Rica. Master's thesis, School of Agronomy, University of Costa Rica, San José.

Laycock, D. H. 1945. Preliminary investigations into the function of endotrophic mycorrhizae of *Theobroma cacao* L. *Tropical Agriculture* (Trinidad) 22:77–80.

Lee, D. W., S. Brammeier, and A. P. Smith. 1987. The selective advantages of anthocyanins in developing leaves of mango and cacao. *Biotropica* 19:40–49.

León, J. 1984. The spread of Amazonian crops in Mesoamerica: The botanical evidence. In *Pre-Columbian plant migration*, ed. D. Stone, 169–73. Cambridge, Mass.: Harvard University Press.

Leston, D. 1970. Entomology of the cocoa farm. *Annual Review of Entomology* 15:273–94.

Levin, D. A. 1971. Plant phenolics: an ecological perspective. *American Naturalist* 10:157–81.

Linnaeus, C. 1753. *Species plantarum*. 2 vols. Stockholm: L. Salvius.

MacLean, J. A. R. 1952. Oil-bearing seeds of possible economic importance to West Africa. *Nature* (London) 307:589–90.

MacLeod, J. J. 1973. *Spanish Central America: A socioeconomic history, 1520–1720*. Berkeley, Calif.: University of California Press.

Malo, F. 1961. Phoresy in *Xenufens* (Hymenoptera: Trichogrammatidae), a parasite of *(Caligo eurilochus)* (Lepidoptera: Nymphalidae). *Journal of Economic Entomology* 54:465–66.

Malo, F., and E. R. Willis. 1961. Life history and biological control of *(Caligo eurilochus)*, a pest of banana. *Journal of Economic Entomology* 54:530–36.

Meeuse, B. J. D. 1975. Thermogenic respiration in aroids. *Annual Review of Plant Physiology* 26:117–26.

Mesler, M. R., J. D. Ackerman, and K. T. Lu. 1980. The effectiveness of fungus gnats as pollinators. *American Journal of Botany* 67:564–67.

Millon, R. F. 1955. When money grew on trees: A study of cacao in ancient Mesoamerica. Ph.D. diss., Columbia University, New York.

Minifie, B. W. 1980. *Chocolate, cocoa and confectionary: Science and technology*. 2d ed. Westport, Conn.: AVI Publishing.

Mori Urpi, J. 1958. Notas sobre el posible origen y la variabilidad del cacao cultivado en América tropical. *Turrialba* (Costa Rica) 8:34–43.

Morris, D. 1882. *Cacao: How to grow and how to cure it.* Kingston, Jamaica.

Myers, J. G. 1930. Notes on wild cacao in Surinam and British Guiana. *Kew Bulletin,* 1–10.

Nagata, K. M. 1985. Early plant introductions in Hawaii. *Hawaiian Journal of History* 19:35–61.

Naipaul, V. S. 1981. *The middle passage.* New York: Random House.

Nosti, R. 1953. *Cacao, café, té.* Madrid: Salvat Editores.

Oviedo, G. F. de. 1851–55. *Historia general y natural de las Indias.* 4 vols. Madrid: Imprenta de la Real Academia de la Historia.

Patino, V. M. 1963. *Plantas cultivadas y animales domesticados en América Equinoccial.* 2 vols. Cali, Colombia: Imprenta Departamental.

Percival, M. 1965. *Floral biology.* London: Pergamon Press.

Pijl, L. van der. 1953. On the flower biology of some plants from Java. *Annales Bogorienses* 1:90–95.

Pittier, H. 1935. Degeneration of cacao through natural hybridization. *Journal of Heredity* 36:385–90.

Posnette, A. F. 1944. Pollination of cacao in Trinidad. *Tropical Agriculture* (Trinidad) 21:115–18.

———. 1945. Incompatibility in Amazon cacao. *Tropical Agriculture* (Trinidad) 22:184–87.

Pound, F. J. 1932a. The principles of cocoa selection. *Agricultural Society of Trinidad and Tobago, Proceedings* 32:122–27.

———. 1932b. Criteria and methods of selection in cacao. *Annual Report on Cacao Research* 2:27–29. Imperial College of Tropical Agriculture, Trinidad.

———. 1933. The genetic constitution of the cacao crop. *Annual Report on Cacao Research* 2:10–24. Imperial College of Tropical Agriculture, Trinidad.

———. 1938. *Cacao and witchbroom disease* (Marasmius perniciosus) *of South America, with notes on other species of* Theobroma. *Report on a visit to Ecuador, the Amazon Valley, and Colombia, April 1937–April 1938.* Port-of-Spain, Trinidad: Yulle's Printerier.

Privat, F. 1979. Les Bromliacées, lieu de developement de quelques insectes pollinisateurs des fleurs de cacao. *Brenesia* (Costa Rica) 16:197–211.

Proctor, M. C. F. 1978. Insect pollination syndromes in an evolutionary and ecosystemic context. In *The pollination of flowers by insects,* ed. A. J. Richards, 105–66. London: Academic Press.

Purseglove, J. W. 1968. *Tropical crops: Dicotyledons.* New York: J. Wiley & Sons.

Ratnam, R. 1961. Introduction of Criollo cocoa in Madras State. *Southern Indian Horticulture* 9:24–29.

Richards, P. W. 1952. *The tropical rain forest: An ecological study*. Cambridge, England: Cambridge University Press.

Roberts, H. H. 1955. New varieties of cocoa in West Africa. In *Cocoa, chocolate, and confectionary alliance: Report on the cocoa conference*, 77–79. London: Cocoa Producers' Alliance and Stephen Austin and Sons.

Roubik, D. W. 1993. Tropical pollinators in the canopy and understory: Field data and theory for stratum "preferences." *Journal of Insect Behavior* 6:659–73.

Roys, R. 1943. *The titles of Ebtun*. Publication No. 505. Washington, D.C.: Carnegie Institution.

Sauer, C. O. 1950. Cultivated plants of South and Central America. In *Handbook of South America Indians*, vol. 6, ed. J. H. Seward, 487–544. Washington, D.C.: Bureau of American Ethnology.

Schultes, R. E. 1958. A synopsis of the genus *(Herrania)*. *Journal of the Arnold Arboretum* 39:217–75.

———. 1984. Amazonian cultigens and their northward and westward migrations in pre-Columbian times. In *Pre-Columbian plant migration*, ed. D. Stone, 32–33. Cambridge, Mass.: Harvard University Press.

Schumann, K. 1886. Vergleichende Bluthenmorphologie der cucullaten Sterculiaceen. *Jahrbeiten Botanische Garten Berlin* 4:286–332.

Simmonds, N. W., 1959. *Bananas*. London: Longmans.

Simpson, B. B., and J. L. Neff. 1981. Floral rewards: Alternatives to pollen and nectar. *Annals of the Missouri Botanical Garden* 68:301–22.

Simpson, B. B., and M. C. Ogorzaly. 1986. *Economic Botany*. New York: McGraw-Hill.

Slobodkin, L. B., and H. L. Sanders. 1969. On the contribution of environmental predictability to species diversity. In *Diversity and stability in ecological systems*, ed. G. M. Woodwell and H. H. Smith, 82–95. Brookhaven Symposia in Biology, no. 22. New York: Brookhaven National Laboratory.

Smythe, N. 1970. Relationships between fruiting seasons and seed dispersal methods in a Neotropical forest. *American Naturalist* 104:25–35.

Soria, J. V. 1959. Apuntes sobre la variabilidad de tipos de cacao en algunas plantaciones de Nicaragua y comentarios sobre su constitución genética. *Cacao* (Costa Rica) 4:1–2.

———. 1966. Principales variedades de cacao cultivadas en América tropical. *Turrialba* (Costa Rica) 16:261–66.

———. 1970a. The latest cocoa expeditions to the Amazon Basin. *Cacao* (Costa Rica) 15:5–15.

———. 1970b. Principal varieties of cocoa cultivated in tropical America. *Cocoa Growers' Bulletin* 15:12–21.

———. 1978. The breeding of cacao *Theobroma cacao* L. *Tropical Agriculture Research Series* 11:161–68. Tropical Agriculture Center, Ministry of Agriculture and Forestry, Japan.

Soria, J. V., F. Ocampo, and G. Paez. 1974. Parental influence of some cacao clones on the yield performance of their progenies. *Turrialba* (Costa Rica) 24:58–65.

Soria, S. de J. 1970. Studies on *Forcipomyia* spp. midges (Diptera, Ceratopogonidae) related to the pollination of *Theobroma cacao* L. Ph.D. diss., University of Wisconsin, Madison.

———. 1975. O papel das abelhas sem ferrao (Meliponinae) na polinizacao do cacaueiro na America tropical. *Revista Theobroma* 5:15–20.

———. 1977a. Dinamica populacional de *Forcipomyia* spp. (Diptera, Ceratopogonidae) na Bahia, Brasil. 2. Variaveis bioticas relacionadas com a polinizacao do cacaueiro. *Revista Theobroma* 7:19–33.

———. 1977b. Dinamica populacional de *Forcipomyia* spp. (Diptera, Ceratopogonidae) na Bahia, Brasil. 3. Variaveis climaticas relacionadas com a polinizacao do cacaueiro. *Revista Theobroma* 7:69–84.

Soria, S. de J., and W. W. Wirth. 1979. Ceratopogonid midges (Diptera: Nematocera) collected from cacao flowers in Palmira, Colombia: An account of their pollinating abilities. *Revista Theobroma* 9:77–84.

Stephenson, A. G. 1981. Flower and fruit abortion: Proximate causes and ultimate functions. *Annual Review of Ecology and Systematics* 12:253–79.

Stone, D. 1984. Pre-Columbian migration of *Theobroma cacao* Linnaeus and *Manihot esculenta* Crantz from northern South America into Mesoamerica: A partially hypothetical view. In *Pre-Columbian plant migration*, ed. D. Stone, 68–83. Cambridge, Mass.: Harvard University Press.

Strand, M. E. 1984. Comparative floral morphology of four *Theobroma* species in relation to pollination biology. Master's diss., University of Wisconsin, Milwaukee.

Swarbrick, J. T. 1965. Estate cocoa in Fernando Po. *Cocoa Growers' Bulletin* 4:14–19.

Taroda, N., and P. E. Gibbs. 1982. Floral biology and breeding system of *Sterculia chicha* St. Hil. (Sterculiaceae). *New Phytologist* 90:735–43.

Taylor, E. L. 1987. Trial by fire: Searching for *Sterculia* in the Amazon. *Harvard Graduate Society Newsletter*, 1–3.

Thompson, J. Eric S. 1956. Notes on the use of cacao in Middle America. *Notes on Middle American Archaeology and Ethnology* 128:95–108. Washington, D.C.: Carnegie Institution.

Torquemada, J. de. 1723. *Monarquía Indiana*. México: S. Chavex Hayhoe.

Toxopeus, H. 1985. Planting material. In *Cocoa*, ed. G. A. R. Wood and R. A. Lass, 80–92. London and New York: Longman.

Urquhart, D. H., and G. A. R. Wood. 1954. Report on a visit to the cocoa zone of Bahia, Brazil. Bournville, Birmingham, England: Cadbury Brothers.

Van Hall, C. J. J. 1914. *Cocoa*. London: MacMillan.

Vogel, S. 1978. Pilzmuckenblumen als Pilzmineten. *Flora* 167:329–98.

Voorhies, B., ed. 1989. *Ancient trade and tribute: Economies of the Soconusco region of Mesoamerica*. Provo: University of Utah Press.

Walker, C. 1959. Notes on the botany of cocoa. *Agricultural Journal of Fiji* 29:56–61.

Wanner, G. A. 1962. *The first cocoa trees in Ghana, 1858–1868*. Basle, Switzerland: Basle Trading Company.

Weaver, M. Porter. 1981. *The Aztecs, Maya, and their predecessors*. New York: Academic Press.

Wellensiek, S. J. 1932. Flower-biological observations with cocoa. *Indonesie* 6:87–101. Archive Koffeicult (read as translation from Dutch).

West, J. A. 1992. A brief history and botany of cacao. In *Chilies to chocolate: Food the Americas gave the world*, ed. N. Foster and L. S. Cordell, 105–21. Tucson and London: University of Arizona Press.

Williams, N. H., and W. M. Whitten. 1983. Orchid floral fragrances and male euglossine bees: Methods and advances in the last sesquidecade. *Biological Bulletin* 164:355–95.

Winder, J. A. 1972. Cacao pollination: Microdiptera of cacao plantations and some of their breeding places. *Bulletin of Entomological Research* 61:651–55.

———. 1977. Some organic substrates which serve as insect breeding sites in Bahian cocoa plantations. *Revista Brasilensis de Biología* 37:351–56.

Winder, J. A., and P. Silva. 1975. Current research on insect pollination of cocoa in Bahia. In *Proceedings of the Fourth International Cocoa Research Conference, Trinidad and Tobago*, 553–65. London: Cocoa Producers' Alliance and Stephen Austin and Sons.

Wirth, W. W. 1956. The heleid midges involved in the pollination of rubber trees in America. *Proceedings of the Washington Entomological Society* 58:241–50.

Wirth, W. W., and W. L. Grogan, Jr. 1988. *The predaceous midges of the world*. Flora and Fauna Handbook No. 4. Leiden: E. J. Brill.

Wirth, W. W., and W. T. Waugh. Five new Neotropical *Dasyhelea* midges (Diptera: Ceratopogonidae) associated with the culture of cocoa. *Studia Entomologica* 19:223–36.

Wolda, H., and C. W. Sabrosky. 1986. Insect visitors to two forms of *Aristolochia pilosa* in Las Cumbres, Panama. *Biotropica* 18:295–99.

Wood, G. A. R. 1991. A history of early cocoa introductions. *Cocoa Growers' Bulletin* 44:7–12.

Wood, G. A. R., and R. A. Lass, eds. 1985. *Cocoa*. London and New York: Longman.

Wright, J. 1907. Cocoa. Ms.

Yamida, M. 1991. Genetic studies in cacao *Theobroma cacao* L. Ph.D. diss., University of Wisconsin, Madison.

Young, A. M. 1981. The ineffectiveness of the stingless bee, *Trigona jaty* (Hymenoptera: Apidae: Meliponinae) as a pollinator of cocoa *(Theobroma cacao)*. *Journal of Applied Ecology* 18:149–55.

———. 1982. Effects of shade cover and availability of midge breeding sites on pollinating midge populations and fruit set in two cocoa farms. *Journal of Applied Ecology* 19:47–63.

———. 1983a. Patterns of distribution and abundance of ants (Hymenoptera: Formicidae) in three Costa Rican cocoa farm localities. *Sociobiology* 8:51–87.

———. 1983b. Patterns of distribution and abundance in small samples of litter inhabiting Orthoptera in some Costa Rican cacao plantations. *Journal of the New York Entomological Society* 91:312–27.

———. 1983c. Seasonal differences in abundance and distribution of cocoa pollinating midges in relation to flowering and fruit-set between sunny and shaded habitats of the La Lola Cocoa Farm in Costa Rica. *Journal of Applied Ecology* 20:801–31.

———. 1983d. *Lirimiris meridionalis* (Schaus), a notodontid moth associated with cocoa (*Theobroma cacao* L.) in Belize. *Journal of the Lepidopterists' Society* 37:182–86.

———. 1984a. *Schizura rustica* (Schaus), a notodontid moth defoliating *Herrania* and *Theobroma* species (Sterculiaceae) in Costa Rica. *Journal of the Lepidopterists' Society* 38:245–49.

———. 1984b. Changes in the distribution and abundance of some Dermaptera between tropical seasons in cocoa plantations at three localities in Costa Rica. *Journal of the New York Entomological Society* 92:184–87.

———. 1984c. Flowering and fruit setting patterns of cocoa trees (*Theobroma cacao* L.) (Sterculiaceae) at three localities in Costa Rica. *Turrialba* (Costa Rica) 34:129–42.

———. 1984d. Ecological notes on cacao-associated midges (Diptera: Ceratopogonidae) in the "Catongo" cacao plantation at Turrialba, Costa Rica. *Proceedings of the Entomological Society of Washington* 86:185–94.

———. 1984e. Mechanism of pollination by Phoridae (Diptera) in some *Herrania* species (Sterculiaceae) in Costa Rica. *Proceedings of the Entomological Society of Washington* 86:503–18.

———. 1985a. Notes on the distribution and abundance of midges (Diptera: Ceratopogonidae and Cecidomyiidae) in some Central American cacao plantations. *Brenesia* (Costa Rica) 24:273–85.

———. 1985b. Pollen collecting by stingless bees on cacao flowers. *Experientia* 41:760–62.

———. 1985c. Studies of cecidomyiid midges (Diptera: Cecidomyiidae) as cocoa pollinators (*Theobroma cacao* L.) in Central America. *Proceedings of the Entomological Society of Washington* 87:49–79.

———. 1985d. Cocoa pollination. *Cocoa Growers' Bulletin* 37:5–23.

———. 1986a. Notes on the distribution and abundance of Dermaptera and Staphylinidae (Coleoptera) in some Costa Rican cacao plantations. *Proceedings of the Entomological Society of Washington* 88:328–43.

———. 1986b. Notes on the distribution and abundance of ground- and arboreal-nesting ants (Hymenoptera: Formicidae) in some Costa Rican cacao habitats. *Proceedings of the Entomological Society of Washington* 88:550–71.

———. 1986c. Distribution and abundance of Diptera in flypaper traps at *Theobroma cacao* L. (Sterculiaceae) flowers in Costa Rican cacao plantations. *Journal of the Kansas Entomological Society* 59:580–87.

———. 1986d. Occurrence of Diptera on tree-trunk mosses in a Costa Rican tropical rain forest. *Pan-Pacific Entomologist* 62:203–8.

———. 1986e. Habitat differences in cocoa tree flowering, fruit-set, and pollinator availability in Costa Rica. *Journal of Tropical Ecology* 2:163–86.

———. 1988a. Population fluctuations of *Azeta versicolor* (Fabricius) (Noctuidae) on *Gliricidia sepium* (Jacq.) (Fabaceae) in northeastern Costa Rica. *Journal of the Lepidopterists' Society* 42:14–18.

———. 1988b. Notes on phenological patterns of flowering and flower-feeding beetles (Coleoptera: Chrysomelidae) in two clones of cacao (Sterculiaceae: *Theobroma cacao* L.) in Costa Rica. *Turrialba* (Costa Rica) 38:143–48.

———. 1989a. Comparative attractiveness of floral fragrance oils of "Rim" and "Catongo" cultivars of cacao (*Theobroma cacao* L.) to Diptera in a Costa Rican cacao plantation. *Turrialba* (Costa Rica) 39:137–42.

———. 1989b. Pollination biology of *Theobroma* and *Herrania* (Sterculiaceae). IV. Major volatile constituents of steam-distilled floral oils as field attractants to cacao-associated midges (Diptera: Cecidomyiidae and Ceratopogonidae) in Costa Rica. *Turrialba* (Costa Rica) 39:454–58.

———. 1991. *Sarapiquí chronicle: A naturalist in Costa Rica*. Washington, D.C.: Smithsonian Institution Press.

Young, A. M., B. J. Erickson, and E. H. Erickson, Jr. 1988. Pollination biology of *Theobroma* and *Herrania* (Sterculiaceae). III. Steam-distilled floral oils of *Theobroma* species as attractants to flying insects in a Costa Rican cocoa plantation. *Insect Science and Its Application* 10:93–398.

Young, A. M., E. H. Erickson, Jr., M. E. Strand, and B. J. Erickson. 1987a. Pollination biology of *Theobroma* and *Herrania* (Sterculiaceae). I. Floral Biology. *Insect Science and Its Application* 8:151–64.

———. 1987b. A trap survey of flying insects in "Finca Experimental La Lola" in Costa Rica. *Turrialba* (Costa Rica) 37:337–56.

———. 1987c. Steam-distilled floral oils of *Theobroma* species (Sterculiaceae) as attractants to flying insects during dry and wet seasons in a Costa Rican cocoa plantation. In *Proceedings, Tenth International Cocoa Research Conference, Santo Domingo, Dominican Republic*, 289–96.

Young, A. M., and A. Muyshondt. 1985. Notes on *Caligo memnon* Felder and *Caligo atreus* Kollar (Lepidoptera: Nymphalidae: Brassolinae) in Costa Rica and El Salvador. *Journal of Research on the Lepidoptera* 24:154–75.

Young, A. M., M. Schaller, and M. Strand. 1984. Floral nectaries and trichomes in relation to pollination in some species of *Theobroma* and *Herrania* (Sterculiaceae). *American Journal of Botany* 71:466–80.

Young, A. M., and D. W. Severson. N.d. Comparative attraction of flying insects to distilled floral oils of cacao cultivars (*Theobroma cacao* L., Sterculiaceae). Ms.

Index

Page numbers followed by *f* indicate an illustration; *t* after a page number indicates a table.

achiote, added to chocolate, 18, 35
Africa: banana cultivation in, 55; cacao cultivation in, 40
Agency for International Development, 64
alkalinization process, 37
Allee, Ralph, 63
Allen, John: on cacao genetics, 70; on cacao tree habitat, 100
Almirante Cacao Research Center, 73
almudes, as cacao seed measure, 31
Alvim, P. de T., on leaf flushing, 91
Amazonia: cacao cultivation in, 11, 34, 38; cacao dispersal in, 6; cacao origin in, 3, 4–8; cacao seed transport in, 9; Great Ice Age impact on, 8–9; wild cacao trees in, 10. *See also* Venezuela
ambergris, added to chocolate, 35
amelonada cacao: characteristics of, 42–43, 43*t*, 44*f*, 45; cultivation of, 38, 40
American Cocoa Research Institute, 62, 64, 114, 121
angoleta cacao, 43, 44*f*
animals: chewing on cacao pods, 11, 97–98, 133–34, 136–37, 158; interactions with cacao trees, 95–99. *See also specific animals*
anise, added to chocolate, 35
annatto (achiote), added to chocolate, 18, 35
Annona muricata, as cacao shade, 23*f*
Anolis, 83
anthocyanins, in cacao leaves, 81, 90
ants, in cacao trees, 95, 97
aphids, in cacao trees, 97
Aphodiplosis, as pollinators, 130
Aphodiplosis triangularis, as pollinators, 149
Apis mellifera scutellata, as pollinators, 113
Arawete Indians, cacao cultivation by, 11
Arias Sánchez, Oscar, 51–52
Aristolochia, pollination of, 132–33

Associated Colleges of the Midwest Costa Rican Field Studies Program, 64
Asurini Indians, cacao cultivation by, 11
Atrichopogon, as pollinators, 120
Aublet, Fusee, 60
Averry, Charles, 63
avocados, 26
Aztecs: cacao cultivation by, 17, 21; cacao seed tribute to, 18

Baker, Dr. James, as chocolate manufacturer, 36
Baker's Chocolate Company, 36
banana cultivation: in Costa Rica, 55–60; pests in, 160–62
banana pseudostems, insects breeding in, 114, 115*f*, 117–19, 122, 123, 168–70
bats: attraction to cacao pulp, 11; cacao distribution and, 10; pod losses from, 46
Bawa, Jamaljit, 128
beans: cacao, *see* cacao seeds; legume, 10, 26
bees, as pollinators, 110–13, 151–52, 166–67
Belize: cacao cultivation in, 17, 21, 24, 73, 134–35; midge populations in, 134–35
Bioko (Fernando Po), cacao cultivation in, 40
birds: cacao distribution and, 10, 85; pod losses from, 46, 98
biting midges, as pollinators, 108, 110, 114–20, 124–30, 134, 146
Bixa orellana, added to chocolate, 18
black pod disease, of cacao, 58
blackwood, as cacao shade, 28
Boston Fruit Company, in banana trade, 57
Bothrops asper (fer-de-lance), 54, 112
Brazil: cacao cultivation in, 34, 38; cacao dispersal in, 6
Bri Bri Indians, *Herrania purpurea* consumption by, 11
bromeliads, 81, 114–16
Brosmium alicastrum, as cacao shade, 23*f*
Bufo toads, in cacao plantations, 121
Bursera simaruba, as cacao shade, 23*f*
butterflies, 54; as banana pests, 57–58; in cacao groves, 160–62

cacao abandonado, 157, 160*f*
cacao bravo, 38
Cacao Ceilan, 71
cacao *común*, 45
Cacao Costa Rica, 71
Cacao Cultivars Register, 72
cacao cultivation, 14–47; abandoned groves of, 157, 160*f*; in Amazonia, 11, 34, 38; in Belize, 134–35; breeding trees for, 101–4; in cenotes, 19*f*, 22, 23, 23*f*, 169; after Conquest, 27–29; in Costa Rica, 25–26, 30, 48–50, 58–60, 73; distribution of, 25*f*, 30–33; earliest, 17–18; *encomienda* system for, 27–29; hedgerows in, 170; irrigation in, 26, 28, 30, 169; in Mesoamerica, 7–8, 18–22; by Pipil-Nicarao peoples, 24–27, 30, 31; plant spacing and arrangement for, 28, 102; pollination enhancement in, 167–71; present status of, 46–47; propagation centers for, 64; rain forests bordering, 157–63; regional differences in, 170; with rubber trees, 61; in small groves, 171–72; in South America, 33–34; spread to other tropical regions, 37–42; start-up costs for, 73; varieties for, 22–24, 26, 42–46
cacao de la tierra (peanuts), 26
cacao de monte, 100
cacao districts, 32
cacao flowers: activity cycles of, 128–29; appearance of, 5*f*, 83–85, 83*f*, 92*f*; buds of, 5*f*; design of, 85–87; fragrance of, 138–52; hermaphroditic, 103; incompatibility phenomenon in, 101–4; opening of, 86–87, 126; pollination of, *see* pollination; protogyny of, 126, 128; seasonal flowering patterns of, 91
cacao fruit. *See* cacao pods
cacao nacional, 45
cacao pods: analogous to heart, 27; animals chewing on, 11, 97–98, 133–34, 136–37, 158; appearance of, 3, 5*f*, 7*f*, 20*f*, 24, 82*f*, 92*f*; characteristics of, 42, 43*t*; fungal diseases of, 40, 58, 68–69, 134, 136–37; growth of, 87; harvesting of, 74; insect breeding in, 96–97, 135–38; of Matina cacao, 159*f*; maturation of, 87; retention on tree, 98–99; ripening of, 7*f*; shapes of, 44*f*; of *T. pentagonum*, 22–23; transport of, 16; uses of, 15–16; wilted, 87; yield of, 46, 67, 74, 137–38
cacao pulp: appearance of, 7*f*, 20*f*; beverage prepared from, 15; fermentation of, 74; as food, 11, 15; as germination inhibitor, 87; nutritional value of, 11; selection for, 3–4
cacao real, 42
cacao seeds: appearance of, 7*f*; arrangement in pod, 20*f*; bitter taste of, 98; characteristics of, 42, 43*t*; chocolate preparation from, 11–12, 15–16, 18, 20; curing of, 76; as currency, 12, 28, 31; defatting of, 36–37; discarding of, 15; dispersal by animals, 98; dry-

ing of, 76, 77*f*, 78*f*; export tax on, 58–59; fermentation of, 74, 76; germination of, 87, 89; in global economy, 33; hybrid, 67–68; liberation of, 11; market prices for, 59–60; in necklaces, 27; planting rituals for, 27; removal from pods, 74, 75*f*; roasting of, 76; as sacred objects, 12–13; selection for, 3; shells of, 78; symbolism of, 27; trading of, 16, 26; transport of, 4, 6, 9, 10; as tribute, 18, 29, 31–32, 33; units of measurement for, 31; uses of, 18, 20; yield of, 31, 67
cacao silvestre, 100
cacao trees: animal interactions with, 95–99; appearance of, 2*f*, 4*f*, 80–81; breeding strategy of, 99–101; cultivation of, *see* cacao cultivation; disease resistant, 47, 68–69; dispersal of, 4–8, 14; domestication of, 10, 17, 23; elevation requirements of, 7; evolution of, 8–9; flowers of, *see* cacao flowers; fungal diseases of, 40, 58, 68–69, 134, 136–37; genetics of, 47, 69–72, 104–6, 164–65; genetic variation of, 8, 44–45; growth of, 89–91; habitat for, 100–101; height of, 80, 89; hybridization of, 34; in Indian art, 16–17, 21; irrigation of, 26, 28, 30, 169; jorquette of, 89, 90*f*; leaves of, 81, 89–91; life span of, 89; origin of, 1–4; pests of, 46; plantation organization of, 12*f*; pods of, *see* cacao pods; rainfall requirements of, 26; rain forest ecological ties of, 157–63, 170, 172; roots of, 92–95; as sacred objects, 11–13; self-compatible forms of, 102; self-incompatibility of, 101–4, 166; shade for, 22, 23*f*, 24, 28, 81, 94, 108, 121; transport of, 42; Trinidad Special Hybrid varieties of, 69; varieties of, 65–69; vegetative propagation of, 67–68, 101; wild. *See also Theobroma cacao;* wild cacao trees
cacau, 15
cachoatl (cacao beans), 18
cacvacentli (cacao fruit), 18
cacvaqualhitl (cacao tree), 18
Cadbury Brothers, as chocolate manufacturer, 36, 37
calabacillos cacao, 45
Caligo, as banana pest, 57–58
Caligo eurilochus, 162
Caligo memnon (owl butterfly), in cacao groves, 160–62
Cameroon, cacao cultivation in, 40
campeche pod, added to chocolate, 35
Canary Islands, banana cultivation in, 55
candy, chocolate, 37
cassava, 24
Castilla tree, 60–61
caterpillars, attacking cacao trees, 95
CATIE (Centro Agronómico Tropical de Investigaciones y Enzeñanza), 61–65, 121
cauliflory, 83–84, 92*f*
Cecidomyiids (gall midges), as pollinators, 108, 110, 130–34, 135, 146
Cecropia, 119
cenotes, cacao cultivation in, 19*f*, 22, 23, 23*f*, 169
Central America, wild cacao trees in, 3
Central American Bank for Economic Integration, 64
Centro Agronómico Tropical de Investigaciones y Enzeñanza, 61–65, 121
CEPLAC (Commissão Executiva do Plano da Lavoura Cacauerira), cacao gene bank of, 70
Ceratopogonidae (biting midges), as pollinators, 108, 110, 114–20, 124–30, 134, 146
Ceylon. *See* Sri Lanka
cherelles, 5*f*, 82*f*; wilting of, 87
Chibcha-speaking Indians, 26
Chichén Itzá, cacao tree depicted at, 19*f*, 21
chile peppers: added to chocolate, 18, 35; cultivation of, 26
China, banana cultivation in, 55
chocolate: allure of, 10–13; archaeologic discoveries concerning, 21; chemical substances in, 13; consumption of, 35–36; flavor of, 79; hot, 35, 156; ingredients added to, 18, 34–35; manufacture of, 35–37, 76–77; milk, 37, 79; origin of name, 20; popularity of, 35; preparation of, 15–16, 18, 20, 34–35; symbolism of, 27; synthetic, 79
chocolate candy, 37
chocolate houses, 35
chocolate liquor, 79
chocolatl (beverage), 18, 20; European impression of, 28–29; history of, 34–35
Choco Province, Columbia, dividing cacao usage practices, 15–16
Chorotega Indians, 25
Christy, Nancy, 150
Chrysophyllum cainito, as cacao shade, 23*f*
chupon, of cacao tree, 82*f*, 89, 90*f*
Churchman, Walter, as chocolate manufacturer, 35–36
cinnamon, added to chocolate, 18, 34
Citrus limonia, as cacao shade, 23*f*
Citrus sinensis, as cacao shade, 23*f*
Clinodiplosis, as pollinators, 131, 133, 137, 149
coca, 26

cocoa, navy, 36
cocoa butter: manufacture of, 36–37, 76; removal of, 79
cocoa powder, manufacture of, 37, 76, 79
Cocoa Research Unit, University of the West Indies, 68, 73
Cocos nucifera, as cacao shade, 23*f*
Codex Mendoza, 18
coffee, 35; cultivation of, 50–54
coffee rust, 51
Colaspis, eating cacao flowers, 97
Colombia: cacao dispersal in, 6, 7, 10; wild cacao trees in, 15
Columbus, Christopher: on cacao seed transport, 49; naming Costa Rica, 55; on rubber, 61
Commissão Executiva do Plano da Lavoura Cacauerira, cacao gene bank of, 70
común cacao, characteristics of, 45
Conquest, cacao cultivation after, 27–29
Copán River Valley, cacao cultivation in, 17
coral tree (*Erythrina*), as cacao shade, 12*f*, 22, 121
Cordia, as cacao shade, 108, 119
corn, 156
Cortés, Hernando, on chocolate beverage, 29, 34
Costa Rica: agricultural research in, 61–73; banana cultivation in, 55–60; cacao cultivation in, 25–26, 30, 48–50, 58–60, 73; cacao dispersal in, 6; cacao transport to, 16; cacao use in, 17; coffee cultivation in, 50–54; railroad building in, 52, 54; rubber cultivation in, 60–63
cotton, 26
Crinipellis perniciosus (witches-broom), 40, 69
criollo cacao: characteristics of, 42–46, 43*t*, 44*f*; cultivation of, 22, 33–34, 38; Java *criollo*, 37; origin of, 2, 5, 14–15; types of, 42
Cuba: banana cultivation in, 57; cacao cultivation in, 34
cultivation: of bananas, 55–60, 160–62; of cacao, *see* cacao cultivation; of coffee, 50–54; of rubber, 60–63; of *Theobroma speciosum*, 11
cundeamor cacao, 43, 44*f*
currency, cacao seeds as, 12, 28, 31

Dasyhelea, as pollinators, 120
Department of Agriculture (Trinidad), 40
Dominican Republic, cacao cultivation in, 34
Dube, Doug, 140, 148
Dutch East India Company, coffee cultivation by, 50
dutching, 37
Dutchman's pipe vine (*Aristolochia*), pollination of, 132–33
Dwarf Cavendish banana variety, 55

East Indies, cacao cultivation in, 40
Ecuador: cacao cultivation in, 45; cacao dispersal in, 7, 10; fungal disease research in, 69; wild cacao trees in, 100, 151
El Salvador: cacao cultivation in, 24–27, 30, 32, 43, 169; cacao yields in, 32
encomienda system, for cacao cultivation, 27–29, 32
England, chocolate houses in, 35
epiphytes, in cacao plantations, 163
Erickson, Barbara: bee pollination studies of, 113; floral fragrance studies of, 140–42
Erickson, Eric H., Jr.: bee pollination studies of, 113; floral fragrance studies of, 140
Erythrina, as cacao shade, 12*f*, 22, 121
Ethiopia, coffee origin in, 50
Euprojoannisia, as pollinators, 108

fanegas, as cacao seed measure, 31
Fannidae, as *Sterculia* pollinators, 84
fer-de-lance, 54, 112
fermentation, of cacao seeds, 74, 76
Fernando Po, cacao cultivation in, 40
Ficus yucatanensis, as cacao shade, 23*f*
Fiji, cacao cultivation in, 42
flowers: cacao, *see* cacao flowers; *Herrania*, 85–87, 88*f*, 129, 138–42
flushing, of cacao tree leaves, 89–91
forastero cacao: characteristics of, 42–46, 43*t*; cultivation of, 38; origin of, 5, 14; pod value of, 67; seed index of, 67; types of, 45
Forcipomyia, as pollinators, 108, 120, 123, 135
Forcipomyia cinctipes, as pollinators, 124
Forcipomyia quatei, as pollinators, 122, 124
Forcipomyia youngi, as pollinators, 122, 124, 125
Fowler, William, on cacao cultivation, 31
fragrance: of *Herrania* flowers, 138–42; of *Theobroma* flowers, 138–52
Frank brothers, in banana trade, 57
Freeman, W. E., cacao-breeding program of, 68
French Plantain banana variety, 55
Fry, Dr. Joseph, as chocolate manufacturer, 36, 37

fungal diseases: of bananas, 58; of cacao, 40, 58, 68–69, 134, 136–37; of coffee, 51
fungus gnats, as pollinators, 109
Fusarium, in cacao pods, 134
Fusarium cabense, in cacao pods, 58

Gage, Rev. Thomas: on cacao tree, 28; on chocolate, 20
Gagné, Ray, 118, 131
Galindo, José, 140
gall midges, as pollinators, 108, 110, 130–35, 146
genetics, of cacao, 3, 8, 44–45, 47, 69–72, 104–6, 164–65
Ghana, cacao cultivation in, 40
Gill & Duffus, in cacao trade, 59
glaciation, forest fragmentation during, 8–9
Gliricidia, as cacao shade, 22, 24, 26
gnats, as pollinators, 109
Goethalsia, 119
Good Neighbor Policy, agricultural research sponsored by, 62
Great Ice Age, forest fragmentation during, 8–9
Gros Michel banana variety, 55
Guardia, Gen. Tomás, on coffee transportation, 52
Guatemala: cacao cultivation in, 17, 21, 24–26, 29–31, 43–44; cacao dispersal in, 6; cacao districts in, 32; cacao yields in, 32; wild cacao trees in, 8
Guiana region: cacao cultivation in, 34; cacao dispersal in, 5
Gulf of Urabá, cacao dispersal in, 6
gum tree (*Heliconia pogonantha*), 162

Hannon, John, as chocolate manufacturer, 36
Hawaii, cacao cultivation in, 41
Heliconia pogonantha, 162
Hemileia vastatrix, 51
hemp, 55
Herrania: breeding strategy of, 100; flowers of, 85–87, 88*f*, 129, 138–42; receptivity to pollination, 126
Herrania purpurea, as food, 11
Hershey Foods Corporation, 21; cacao farms of, 73, 134–35
Hevea brasiliensis, cultivation of, 60–61
Honduras: cacao cultivation in, 17–18, 24, 43; cacao dispersal in, 6
honey, added to chocolate, 20
hot chocolate, 35, 156
"humble cacao," 24
Humboldt, Alexander von: on cacao harvests, 46; on cacao productivity, 170; on cacao products use, 15–16
Hummingbird Hershey Cocoa Farm, 73, 134–35
Hunter, J. Robert: on cacao pod numbers, 137; research of, 63–64; rubber trees planted by, 61
hybridization, of cacao, 67–68

Imle, Dr. Ernest P., as Rubber Institute head, 62–63
Imperial College of Tropical Agriculture (Trinidad), 40
India, cacao cultivation in, 38, 42
Indonesia, cacao cultivation in, 41
Inga, as cacao shade, 22
insects: attacking cacao trees, 95–96; attracted to floral fragrances, 143–47; in cacao plantations, 81. *See also specific insect*
Institute of Interamerican Affairs, 63
Instituto Interamericano de Ciencias Agrícolas, 62–63
International Cocoa Gene Bank, 70
irrigation, of cacao plantations, 26, 28, 30, 169
Isabel II of Spain, 40
Izalco, El Salvador, cacao cultivation in, 30–31, 32

jaguars, 49, 54, 156
Jamaica: banana cultivation in, 57; cacao cultivation in, 34
Janos, David P., on mycorrhizal activity, 93–95
Java: cacao cultivation in, 38; coffee cultivation in, 50
Java *criollo* cacao, 37
jorquette, of cacao tree, 89, 90*f*

Keith, Minor, in banana trade, 57
King Ranch, cattle farming research sponsor, 63

Lacandon cacao, 22, 23
lagarto cacao, 22, 42
La Lola Experimental Farm, 64, 66*f*
laurel, as cacao shade, 108
leaf beetles, attacking cacao trees, 95
leaf litter. *See* mulch layer

leaves, of cacao trees, 81, 89–91
Le Lacheur, William, as coffee trader, 51
Lesser Antilles, cacao cultivation in, 34
Lewontin, Richard C., 115
Linnaeus, Sterculiaceae description by, 85
lizards, 83
London Cocoa Trade Amazon Project, 8, 47, 70
Lord Cacao, 21
Lynne, Paul, 148

MacArthur, Robert, 115
McPhail trap, for floral fragrances, 143–44, 145*f*
Madagascar, cacao cultivation in, 42
madre de cacao, 24, 28
maize, 26; added to chocolate, 18
Malaysia: cacao cultivation in, 60; rubber cultivation in, 61
Maldonado, Alonso, 29
Mangifera indica, as cacao shade, 23*f*
manila hemp, 55
manu, as cacao shade, 108
Marin, Francisco de Paula, 41
Martinique, cacao cultivation in, 34, 38
Matina cacao, 71, 158; characteristics of, 45; fruits of, 159*f*; insect populations of, 136–37; pollination of, 153
Matlick, B. K., 134
Maxillaria, 81
Maya Indians: cacao cultivation by, 17, 21–24, 163, 167, 169; cacao importance to, 18; cacao selection by, 71; cacao types recognized by, 24; chocolate use by, 20; view of natural world, 21
Mediterranean region, banana cultivation in, 55
Megaselia, as pollinators, 133
Melicoccus bijugatus, as cacao shade, 23*f*
Mesoamerica/Mexico: cacao cultivation in, 7–8; cacao dispersal in, 6; cacao domestication in, 17; cacao revered in, 11–13; chocolate use in, 15–16; wild cacao trees in, 8
methylxanthine, 13
Mexico. *See* Mesoamerica/Mexico
Meyen, Franz Julius Ferdinand, 41
midges: in ancient cultivation sites, 167–69; attraction to floral fragrances, 143–52; in Belize, 134–35; biting, 108, 110, 114–20, 124–30, 134, 146; breeding sites for, 114–24, 135–38, 168, 170; in cacao pods, 97; daily activity periods of, 126, 128–29; dispersal tendencies of, 165–66, 167; gall, 108, 110, 130–34, 135, 146; in plantations, 165, 167, 168; pollination technique of, 127*f*, 132–33; in wild cacao clumps, 164
milk, added to chocolate, 35
milk chocolate, 37, 79
Milwaukee Public Museum, 118–19
Minquartia guinanensis, as cacao shade, 108
M&M Mars Company, cacao farms of, 73
mole sauce, 18
molinillo, 20
Monilia, as cacao disease, 136–37
Moniliophthora, as cacao disease, 59
Moniliophthora roreri, as cacao disease, 69
monkey(s), 54; attraction to cacao pulp, 11; cacao distribution and, 10; eating cacao pods, 97–98; pod losses from, 46; *Sterculia* seed dispersion by, 85
monkey ladder vines, 54
Monolepta, eating cacao flowers, 97
Montezuma, 21, 34
Morgan, Henry, 50
Mueller, Ludwig, 140
mulch layer: in cacao plantations, 81, 92–95, 114–19, 168–70; in rain forest, 163
Musa acuminata, 55
Musa balbisiana, 55
Musa paradisiaca, 55; as cacao shade, 23*f*
Musa sapientum, 55
Musa textilis, 55
Muscidae, as *Sterculia* pollinators, 84
Mycetophilidae, as pollinators, 109
Mycodiplosis, as pollinators, 131, 133, 134
Mycodiplosis ligulata, as pollinators, 149
Mycodiplosis ligulata Gagné, as pollinators, 131–33
Mycorrhizae, interactions with roots, 92–95

Naipaul, V. S., describing Trinidad cacao plantation, 81
navy cocoa, 36
nibs: defatting of, 36–37; grinding of, 79
Nicaragua: cacao cultivation in, 24–28, 31, 43; cacao dispersal in, 6; cacao pulp use in, 15
Nicarao Indians, cacao cultivation by, 24–27
Nigeria, cacao cultivation in, 40
Northern Railway (Costa Rica), 63
nuts, added to chocolate, 35

Olmec Indians, cacao cultivation by, 17
orange water, added to chocolate, 35
orchids, 18, 81

Organization for Tropical Studies, 64
Orinoco River region: cacao dispersal in, 6; cacao origin in, 3
Oviedo y Valdes, Gonzalo Fernández de, on cacao cultivation, 27–28, 31
owl butterfly: as banana pest, 57–58; in cacao groves, 160–62

Panama: banana cultivation in, 57; cacao dispersal in, 6; cacao transport to, 16
Panama disease, of banana, 58, 59
Pará cacao, 45
Pará rubber, 60–61, 108
parasites, of *Caligo* caterpillers, 162
parrots, 158; eating cacao pods, 46, 98; *Sterculia* seed dispersion by, 85
pataxte, 24, 26
Patterson, Gordon, 134
peanuts, 26
pentadecene, in cacao flower fragrance, 142
peppers, 156
Peter, Daniel, as milk chocolate inventor, 37
phenylethylamine, 13
Philippines, cacao cultivation in, 37–38
phorid flies, as pollinators, 110, 129–30
Phytophthora, as cacao disease, 58, 134, 136–37
pinolillo, 15
Pipil Indians, cacao cultivation by, 24–27, 30, 31
plantain, wild, 57
pods, cacao. *See* cacao pods
pod value, 67
pollen, cacao flower, 86, 112; sticky surface of, 130; viability of, 126
pollen thieving, 112, 151–52
pollen tubes, growth of, 128
pollination (cacao flower), 107–54, 157–58; bees in, 110–13, 151–52, 166–67; "buzz," 112; enhancement of, 167–71; field studies of, 124–30; flower receptivity to, 126; flower structure and, 83–84; fragrance and, 138–52; hand, 67–68, 108; midges in, *see* midges; in plantations, 164; pollinator breeding sites and, 114–24, 135–38; self-compatibility and, 101–4; time of, 128–29; of wild trees, 164–67
pollination (*Sterculia*), 84–85
Polynesia, banana cultivation in, 55
Posnette, A. F., on cacao self-incompatibility, 102
potatoes, 10
Pound, F. J., cacao research by, 67, 69, 103
Pourouma, 119
Pouteria mammosa, as cacao shade, 23*f*
Príncipe, cacao cultivation in, 40
protogyny, of cacao flowers, 126, 128
pulp, cacao. *See* cacao pulp

Quaternary period, forest fragmentation during, 8–9

rain forest, cacao groves bordering, 157–63, 170, 172
Ramírez, William, 113
rats, attraction to cacao pulp, 11
refuge theory, 9
Regional Rubber Institute, 62–63
Richards, P. W., on cacao flowers, 83–84
Ridley, Henry, as rubber expert, 61
rodents, cacao distribution and, 10
roots, of cacao trees, 92–95
Rowntree family, as chocolate manufacturer, 36
rubber trees: as cacao shade, 81; cultivation in Costa Rica, 60–63
rum, added to chocolate, 36

Sabal yapa, as cacao shade, 23*f*
Saint Lucia, cacao cultivation in, 38, 73
Samoa, cacao cultivation in, 42
Samper, Armando, 63
Sánchez Lepis, Julio, as coffee cultivator, 51
Sanders, Nicholas, mixing chocolate with milk, 35
Santo Domingo, banana cultivation in, 55, 57
São Tomé, cacao cultivation in, 40
Sarcophagidae, as *Sterculia* pollinators, 84
Scavinia cacao variety, 69
Schaller, Marilyn, 138
Schultes, Richard: on cacao spread, 5–7; on rubber, 60
Sciaridae, as pollinators, 109
seed index, 67
seeds, cacao. *See* cacao seeds
Servicio Técnico Interamericano di Cooperación Agrícola (STICA), 64
Severson, David, 147
Sewall Wright effect, 164
shade, for cacao trees, 22, 23*f*, 24, 26, 28, 81, 94, 108, 121
Silk Fig banana variety, 55
Singapore, cacao cultivation in, 42
Siparuna, pollination of, 134

Sloane, Sir Hans, on hot chocolate, 35
sloth, 158
Smithsonian Institution, 118–19
Soconusco region, cacao cultivation in, 18
soils, mycorrhizal activity in, 92–95
Sonsonate region, cacao trading in, 32
Soria, Saulo de Jesus, 114
soursop (*Annona muricata*), 23*f*
Spain, chocolate introduction to, 34–35
Spaniards: cacao cultivation by, 27–30, 32–34, 49–50; cacao transfer by, 37
spider webs, midges on, 130–31
Spivak, Maria, 113
squash, 10, 26, 156
squirrels: attraction to cacao pulp, 11; opening cacao pods, 46, 97, 133–34, 136–37
Sri Lanka: cacao cultivation in, 38, 41–42; rubber cultivation in, 61
Standard Fruit Company, banana cultivation by, 59
Sterculia, pollination of, 84–85
Sterculia chicha, flowers of, 84
STICA (Servicio Técnico Interamericano di Cooperación Agrícola), 64
Strand, Melanie, 126, 131, 138, 139
Suchard chocolate, 37
sugar, added to chocolate, 35
Sulawesi, cacao cultivation in, 38, 41
Sumatra, cacao cultivation in, 38
Sweetheart Cup Corporation, 115

Tanzania, cacao cultivation in, 42
tapioca, 24
Taylorb, Orley, 113
tea, 35
Theilaviopsis, in cacao pods, 134
Theobroma angustifolium, cultivation of, 22, 26
Theobroma bicolor, 24; beverage from, 15; cultivation of, 22, 26; pods dropping from, 99; seeds of, 28
Theobroma cacao: in cenote, 23*f*; cultivation of, 26; dispersal of, 4–8; flowers of, 138–52; fruit of, 82*f*; glaciation impact on, 9; origin of, 1–4; subspecies of, 14. *See also criollo* cacao; *forastero* cacao
Theobroma cacao subsp. *cacao*: distribution of, 6*f*; domestication of, 24
Theobroma cacao subsp. *cacao* form *lacandonense* Cuatrecasas, 22
Theobroma cacao subsp. *sphaerocarpum*, 45; distribution of, 6*f*
Theobroma grandiflorum, flowers of, 84
Theobroma leiocarpa, 3
Theobroma mammosum, 46; flower fragrance of, 139, 142; pollination of, 132
Theobroma pentagona, 3, 22–23, 42
Theobroma simiarum, 46; flower fragrance of, 142; pollination of, 132
Theobroma speciosum: cultivation of, 11; flower fragrance of, 139, 142
Theobroma speciosum Willd, 86*f*
theobromine, 13; in cacao pulp, 3–4; extraction of, 78
tlacachuaquahuitl, 24
toads, in cacao plantations, 121
tobacco, 26
tomatoes, 26, 156
toucans, 158
Trema, 119
Trigona, as pollinators, 110–13
Trinidad: cacao cultivation in, 34, 40, 42, 72–73; cacao research in, 68
Trinidad Special Hybrid varieties, of cacao, 69
trinitario cacao, 34; characteristics of, 42–43, 43*t*, 46; cultivation of, 40; pod value of, 67; seed index of, 67

Union Vale Cocoa Estate, 73
United Fruit Company, 64; in banana trade, 57; cacao cultivation by, 58; cacao varieties developed by, 66–67
United States Chocolate Manufacturers Association, 74
United States Department of Agriculture, cacao gene bank of, 70
University of the West Indies, Cocoa Research Unit, 68, 73

van Hooten, Conrad, chocolate process of, 36–37
vanilla, added to chocolate, 18, 35
Vásquez de Coronado, Juan, 49
vegetative propagation, of cacao trees, 67–68, 101
Venezuela: cacao cultivation in, 33–34; cacao dispersal in, 5, 6, 7; cacao transport from, 16; cacao uses in, 16; wild cacao trees in, 15, 34
von Storen, José Heinrich, 61

Walter Baker and Company, Ltd., 36
Wardian cases, for cacao tree transport, 42
Warneau, Frederick, 38
West Indies, cacao cultivation in, 36

Wickham, Harry, as rubber expert, 61
wild cacao trees: breeding strategy of, 99–101; Central American population of, 14; clumping of, 99; dispersal of, 10; distribution of, 2–3, 6*f*, 8, 100; domestication of, 10; genetics of, 3, 165; growth patterns of, 100–101; habitat for, 100–101, 163–64; harvesting from, 10; hybridization with domesticated types, 47; insect breeding in, 135–38; intercropping with, 24; mycorrhizal colonization of, 94; pollination of, 151, 164–67; reproduction of, 163–64; vegetative propagation of, 101; in Venezuela, 15, 34
wild plantain, 57
Williams, Norris H., 139–40
Winder, John, 114
Wirth, Bill, 118, 122, 123
witches-broom, 40, 69
World's Finest Chocolate Company, cacao farms of, 73

Xanthosoma yucatanense, as cacao shade, 23*f*
xiquipil, as cacao seed measure, 31

yaquaquyt, as cacao shade, 28
yuca, 24, 26
Yucatán region, cacao cultivation in, 17, 22